Atlas of Ultrastructure Interaction Proteome Between Barley Yellow Dwarf Virus and Gold Nanoparticles

Authored by
Noorah Abdulaziz Othman Alkubaisi
and Nagwa Mohammed Amin Aref

Published in London, United Kingdom

IntechOpen

Supporting open minds since 2005

Atlas of Ultrastructure Interaction Proteome Between Barley Yellow Dwarf Virus and Gold Nanoparticles
http://dx.doi.org/10.5772/intechopen.95313
Authored by Noorah Abdulaziz Othman Alkubaisi and Nagwa Mohammed Amin Aref

Contributors
Noorah Abdulaziz Othman Alkubaisi, Nagwa Mohammed Amin Aref

First published in London, United Kingdom, 2021 by IntechOpen
IntechOpen is the global imprint of INTECHOPEN LIMITED, registered in England and Wales, registration number: 11086078, 5 Princes Gate Court, London, SW7 2QJ, United Kingdom
Printed in Croatia

British Library Cataloguing-in-Publication Data
A catalogue record for this book is available from the British Library

Additional hard and PDF copies can be obtained from orders@intechopen.com

Atlas of Ultrastructure Interaction Proteome Between Barley Yellow Dwarf Virus and Gold Nanoparticles
Authored by Noorah Abdulaziz Othman Alkubaisi and Nagwa Mohammed Amin Aref
p. cm.
Print ISBN 978-1-83969-433-2
Online ISBN 978-1-83969-434-9
eBook (PDF) ISBN 978-1-83969-435-6

Meet the authors

Noorah Abdulaziz Othman Alkubaisi has been working in the field of plant virology, virus immunology, botany, and microbiology since 2013. She holds a degree from the College of Science, King Saud University, Saudi Arabia. She is a senior scientist in a research group of the Department of Botany and Microbiology Lab. She has been Assistant Vice Dean of Student Affairs at King Saud University since 2019. From 2017 to 2020, she was assistant director of the Innovation and Entrepreneurship Unit at the same university. She is the recipient of several awards including the best global innovator of KIPA Award in 2019, a bronze medal among 720 international inventions at the Seoul International Invention Fair, Republic of Korea, in 2019, the Innovation Award from King Abdulaziz University, Jeddah, in 2019, and the Excellence Award for the best faculty member at the college level in 2018. Additionally, she is a holder of a patent entitled "Method of Inhibiting plant virus using gold nanoparticles."

Over the past decade, Prof. Nagwa Mohamed Mohamed Amin Aref has worked in the areas of virology, molecular virology, nanotechnology, and immunology. Her work as an inventor has investigated three applied patents methods: a method of treating a bacterial infection using colostrum, a method of using clay suspension to prevent viral and phytoplasma diseases in plants, and a method of inhibiting plant virus using gold nanoparticles. She has made fundamental contributions to plant and medical viruses including publishing more than ninety-one journal articles and presenting at numerous conferences. Included topics are: "Detecting plant viral disease of stone fruit trees", "Engineering transgenic tomato plants Resistant to Tomato Yellow Mosaic Gemini Virus", "Radioprotective efficacy of zinc oxide nanoparticles on γ-ray-induced nuclear DNA damage in *Vicia faba* L. as evaluated by DNA bioassays", "Modulatory effect of zinc oxide nanoparticles on gamma radiation-induced genotoxicity in *Vicia faba*", "Bioactive Molecules from Dodder *Cuscuta* as a critical parameter in the management of plant virus disease", "Evaluation o Biological and Molecular Characterizations for Identification of a Phytoplasma Associated With Lemon Witches'-Broom in Egypt", "Physiological parameters correlated with Tomato Mosaic Virus inducing a defensive response in Datura metel", "Interleukin 17 Level as a Prognostic Marker in Highly Active Antiviral Treated Human Immunodeficiency Virus (HIV) in Saudi Patients", "Lymphocyte subset and anti-HLA in AIDS Saudi patients", "Prevalence of HCV Genotypes and Viral Load in Saudi Arabia", "Correlation Between Phage Typing and Toxins Content as an Outbreak Tool in *Staphylococcus aureus*", "Olive Leaf Extract Trigger Defense Physiological Markers in *Datura metel* against Tobacco Mosaic Virus" "Correlation between Hepatitis B Surface Antigen Titers and Hepatitis B Virus DNA Levels" and etc. Over 91 journal articles have been published in these areas, and also numerous conference presentations made. She has collaborated with US, Indian, Saudi, Egyptian, German, French, and Taiwanese researchers. Previously, she was Professor of Virology, Molecular Virology, Faculty of Agriculture, Microbiology Department, Ain Shams University, Cairo, Egypt.

Contents

Preface

Nanotechnology in agriculture is at its nascent stage. A recent remarkable innovation at King Saud University, College of Science, Saudi Arabia—proteome platforms—provides crucial resources to promote research in gold nanoparticle (AuNP) applications in virus-infected plants. This book examines AuNP behavior via transmission electron microscopy (TEM) inside plant-infected cells, as there is a lack of knowledge in the field of virus pathogenicity at the level of ultrastructure. TEM is a powerful tool in elucidating plant cells' fine details at the nanoscale. The book describes each organelle's structure of plant cells over the following six chapters: "Introductory Chapter: Atlas of Ultrastructure Interaction Proteome between Barley Yellow Dwarf Virus and Gold Nanoparticles"; Chapter 2, "Nucleus"; Chapter 3, "Chloroplast and Mitochondria"; Chapter 4, "Cell Wall"; Chapter 5, "Cytoplasmic Matrix and Viroplasms Inclusions in the Presence of Gold Nanoparticles (AuNPs)"; and Chapter 6, "The Intervention of Gold Nanoparticles (AuNPs) Interactions Lead to the Disappearing of Virus Particles" The applied plant species is Hordeum vulgare (Barley), and we selected the most severe barley yellow dwarf virus (BYDV) isolates. The most compelling moment of plant virus entry considering is inducing viral infection crossing the cell-wall barriers. We knocked out and interfered with the virus as bionanoparticles with other metal AuNPs on the plant. A combinatorial approach using the integration of virus with AuNPs in our proteome platform, like vaccination, is now an effective strategy with dual effects of protecting plants and ruining virus particles inside the infected cells. Application of AuNPs in the presence of virus infection encourages the plants to produce reactive oxygen species (ROS) continually in structures such as chloroplasts, mitochondria, peroxisomes, the endoplasmic reticulum (ER), and plasma membranes. These resemble components of the defensive system.

The authors would like to thank the KSUcentral lab-Electron Microscopy Unit for their efforts and financial support.

Noorah Abdulaziz Othman Alkubaisi
Department of Botany and Microbiology,
College of Science,
King Saud University,
Riyadh, Kingdom of Saudi Arabia

Nagwa Mohammed Amin Aref
Department of Microbiology,
College of Agriculture,
Ain Shams University,
Cairo, Egypt

Chapter 1

Introductory Chapter: Atlas of Ultrastructure Interaction Proteome between Barley Yellow Dwarf Virus and Gold Nanoparticles

Noorah Abdulaziz Othman Alkubaisi and Nagwa Mohammed Amin Aref

1. Introduction

Interaction between Barley yellow dwarf virus, BYDV-PAV, and gold nanoparticles AuNPs application revealed great effect whether in *vitro* or *Vivo.* The significant effect of virus particles occurred inside the plant cell due to the existence of AuNPs treatment. It was clear that using tiny AuNPs 3.151 to 31.67 nm had a potential agent to ruined virus particles inside the infected cells. AuNPs cause damage to the virus-like particles (VLPs) of the barley yellow dwarf virus-PAV. Where they observed puffed and deteriorated VLPs decorated with AuNPs, as well as destroyed and vanished particles, using Transmission Electron Microscopy TEM. Generally, the plant cell contained different organelles that exhibited ultrastructure changes in Nucleus, Chloroplast, Plant cell wall, Mitochondria, cytoplasmic matrix, and viable cellular composition of the infected cell with AuNPs. TEM is a powerful tool in elucidating plant cells' fine details at the nanoscale. The present Atlas describes each organelle's structure of plant cells revealed by TEM in healthy, infected, and treated with AuNPs in **Figure 1**.

The purpose of this work is expressed via TEM, which is a very accurate tool for judging the AuNPs behavior inside the plant infected cell. Recent remarkable innovations in KSU. Platforms [1, 2] provide crucial resources to promote research in AuNPs applications and applied plant species as *Hordeum vulgare* (Barley) due to lacking knowledge in the field of virus pathogenicity at the level of ultrastructure. A combinatorial approach using the integration of virus with AuNPs in our proteome platform is now an effective strategy for clarifying molecular systems integral to improving plant productivity of the frequency and importance of ultrastructure abnormalities in Barley crop development caused by BYDV-PAV. It was finding out the critical feedback of using AuNPs applications on plant cells by examining the virus's behavior conjugated with AuNPs by ultra-structure in TEM on the diseased plants.

The method of inhibiting a plant virus using AuNPs is a method of inducing plant resistance against viral disease caused by BYDV by introducing a thoracically adequate amount of polydispersed AuNPs system integrated with the virus particles wherein virus particles were dissolved and melted in **Figure 2**. The application of

IntechOpen

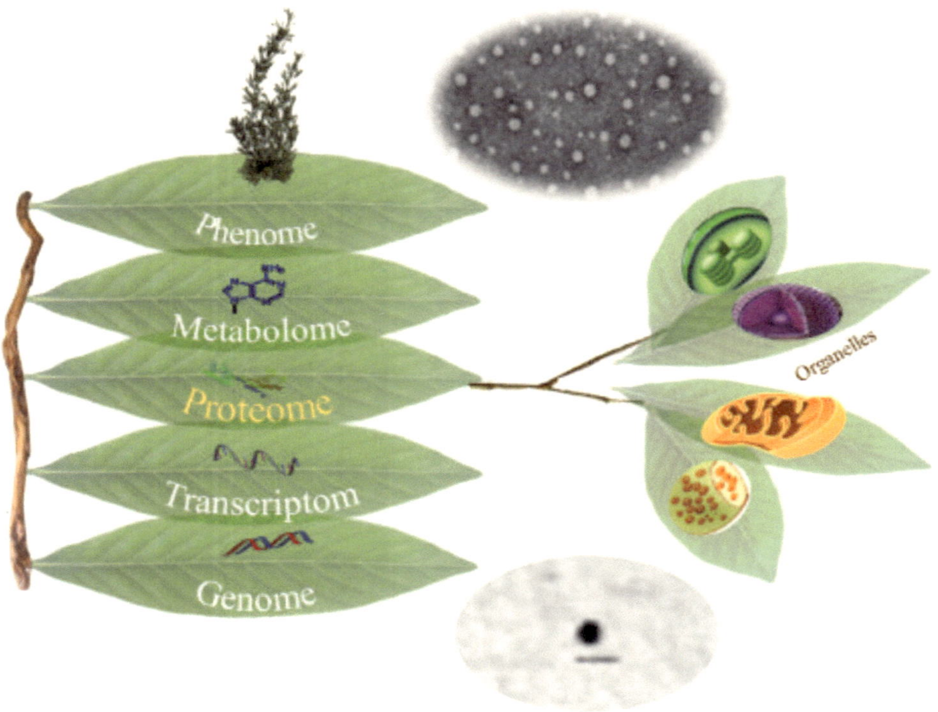

Figure 1.
The plant cell contained different organelles that exhibited ultrastructure changes in nucleus, chloroplast, plant cell wall, and mitochondria that describes each of them by TEM in healthy, infected, and treated with (AuNPs) as interaction preteome of BYDV- PAV.

nanotechnology in agriculture, even at its global level, is at its nascent stage. The most crucial moment of plant virus entry considering is inducing viral infection crossing the cell/wall barriers. In our study, we knocked out and interfered with the virus as bio Nanoparticles with other metal AuNPs on the plant; [3] we selected the most severe BYDV isolates for Nano application on the plants [4].

The studied BYDV is the type member of the luteovirus group [5]. The Latin"luteo" name means yellow [6] and describes the most typical infected plants by luteoviruses. BYDV considered a model for dealing with the "yellowing" virus diseases, as reported [7]. Most infections appear as necrosis in the phloem, which leads to external symptoms such as stunning and leaf chlorosis [5]. The exact symptoms were reported [8] that might cup inward, tender, and show more stiffness than usual. BYDV is spread by aphids and induces the most widely and most distributed and most destructive virus disease globally.

2. Gold nanoparticles have adual positive effect

The prospect antiviral characteristic of Metal nanoparticles (MeNPs) in nano-agriculture drive them as a potential factor for commanding these histological agents. It is essential to detect the dosage of NPs, the application intervals, their effect as a biostimulant. The clarification of the mechanisms of action, are not fully understood [9]. Application of AuNPs in the presence of virus infection encourages the plants to produce Reactive Oxygen Species (ROS) continually in structures such as chloroplasts, mitochondria, peroxisomes, the endoplasmic reticulum (ER),

DOI: http://dx.doi.org/10.5772/intechopen.97438

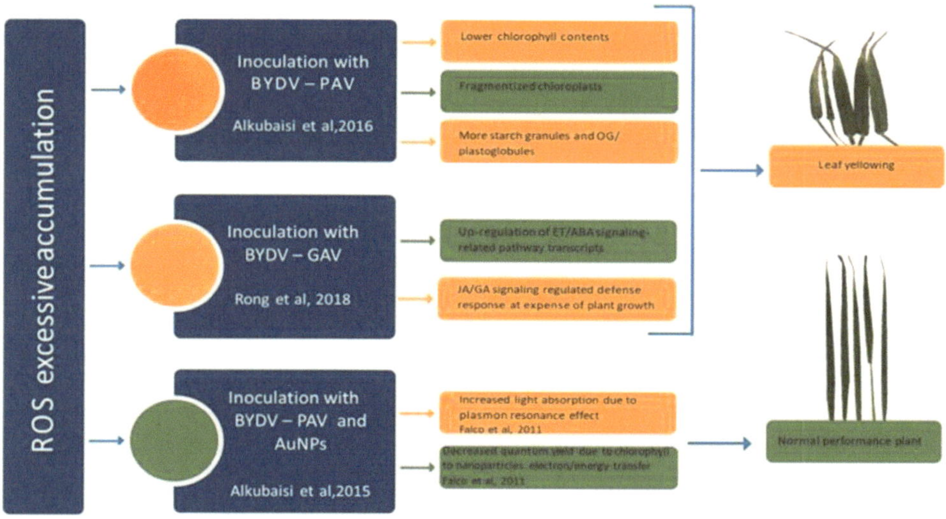

Figure 2.
Methodology, pathological alternation of the plant cellular components, and the dual beneficial effect of AuNPs on the plant performance through the following treatments in the electron micrographs; I. heathy cells from barley leaves, II. Infected cells from by BYDV- PAV from barley leaves, III. Pretreated leaves of barley with AuNPs and infected by BYDV-PAV.

and plasma membranes [10]. These resemble components of the defensive system that have been classified according to their catalytic activity, molecular weight, the compartment where they act, and level of defense or mechanism of action [11]. The positive effect of AuNPs, therefore, needs further study to explore the physiological and molecular mechanisms. However, due to the tiny size, reactivity, and efficient penetration ability, metal nanoparticles could reach many intracellular and extracellular plant sites. That may trigger a set of physiological processes such as senescence affecting plant growth, crop yield, and ecological productivity [12]. Nanoparticles (NPs) have unique physicochemical properties, i.e., high surface area, high reactivity, tunable pore size, and particle morphology. The appropriate elucidation of the physiological, biochemical, and molecular mechanisms of nanoparticles in plants leads to better plant growth and development [13]. The chemical reactions, especially reduction–oxidation reactions, were catalyzed by Ag, Au, Fe, and Co. The released nano ions may alter proteins while entering into cells. Mechanical effects rely on the size of nanoparticles [14]. For example, the cell wall damage can be caused by the high concentration adsorption of hydrophobic nanoparticle retention and may cause clog pores, which can interfere with water uptake [15]. The ability to pass through the cell wall might not be a prerequisite for causing oxidative stress and toxicity. Some researchers suggest that despite nanomaterials' inability to pass through plants' cell walls, they can cause oxidative stress and eventually lead to chromosome condensation [16].

Similarly, CuO nanoparticles can also cause oxidative damage to plant DNA and can be detected in plant cells [17]. Particles with anoxic surface often form a layer of OH– groups at the surface; these negatively charged groups attract positively charged side groups of proteins [14]. Surface effects have engaged a great deal of attention in the field of nanotoxicology.

The positive, neutral, and negative charge of Au nanoparticles are resemble hydroponic exposure to Rice plants. The distribution of the bioaccumulated Au nanoparticles due to the negative surface charge of the nanoparticles, which is more toxic in above-ground organs [18]. The small particle modulation of 15 nm or 25 nm AuNPs was transported to the shoot in poplar. At the same time, larger particles (50 nm) could hold their size in vivo. AuNPs were located within the roots in a large amount than in leaves. AuNPs were detected in different tissues, phloem complex, xylem, cell wall, plastids, mitochondria, and more abundantly in the plasmodesmata [12]. Negatively charged, anionic carboxylate AuNPs conferred protection to the model lipid membrane against the extreme pH (=12) via shielding effects, whereas positively charged, cationic amino-AuNPs could penetrate and disrupt the model membrane [19].

Nanotoxicity is based on empirical data by an exact predictive model, which is explained by the interaction between surface charge and particle size that affects AgNPs toxicity in both the prokaryotic and eukaryotic model organisms [20]. The interaction between particle size and potential surface charge influencing ENM phytotoxicity has not received much attention. Therefore, the potential effect surface charge density remains to be tested in plants. ORF3 encodes a major coat protein (CP) of 22 kDa [21]. The coat protein plays a crucial role in maintaining a high level of accumulation of genomic RNA, though unnecessary for PAV replication [22]. ORF4 is entirely nested within ORF3 and codes for a 17 kDa nonstructural protein required for BYDV-PAV to spread systemically in plants [23]. The expression of ORF 4 is associated with a unique regulatory mechanism of the ribosome leaky scanning mechanism [24]. The ORF4 translation product is similar to that of the homology of ORF4 in potato leafroll virus (PLRV), which has biochemical properties specific to known movement proteins, including the ability to be phosphorylated, binding nonspecifically to nucleic acids [25, 26] and localization to the plasmodesmata [27]. PAV ORF5 is fused to CP as a readthrough domain and encodes a 50 kDa protein expressed as a 72 kDa fusion protein via a readthrough suppression of the ORF3 stop codon [23, 28–30]. A frameshift mutation within ORF6 was reported to be incompatible with BYDV-PAV RNA replication in protoplasts [31]. In [32], it is found that the RNA sequence encoding or flanking ORF6, rather than the protein product of ORF6, is required for PAV replication in oat protoplasts [33].

The viral infection starts with virus replication in the infected cell initially and spread to neighboring cells through plasmodesmata which are considered intercellular conduit connecting cell walls. This process is called cell-to-cell (short-distance) movement, facilitated by viral movement protein (MP). The following phase is termed (long-distance) movement which viruses could enter the vascular tissue, dispense, and flood into non-infected tissues, helped by the phloem stream [34]. It is presumed that the cell-to-cell movement is an active function, requiring specific interaction between the virus and plasmodesmata, whereas systemic viral spread through the vascular tissue is a passive process, driven by the flow of photoassimilates [35]. The discovery that a 30 kDa movement protein (MP) encoded by the Tobacco mosaic virus (TMV) was required for viral cell-to-cell movement [36, 37] exploring the trafficking mechanisms of a more comprehensive viral array opened a new path. Trafficking viral protein and RNA by viral MP into the phloem and their inter-organ regulation of plant development were rarely studied for some viruses [38] and [39]. The viral nucleic acid conjugate with the MP, which could transport it through plasmodesmata. The first viral MP of the Tobacco Mosaic Virus (TMV) was discovered that had 30 kDa protein (P30) and was able to bind single-stranded nucleic acid [40], mediated by two independently active domains of the MP [41]. The P30-TMV RNA complex measures a diameter of 1.5–3.5 nm [41] and [42] and may interact with the cytoskeletal elements to facilitate the transport of the P30- TMV RNA complex from cytoplasm to plasmodesmata [43] and [44]. The diameter is even smaller than a protein-free, folded TMV RNA, allowing easy access through dilated plasmodesmata [45, 46]. The MP could bind nonspecifically to single-stranded RNA and DNA in vitro [25] and associate with plasmodesmata in host plants [47]. ORF 4 proteins in luteoviruses may provide a clue for assisting virus cell-to-cell spread in host plants [48] as there is a high similarity of amino acid sequence between ORF 4 protein encoded by luteoviruses and PLRV MP [49, 50]. BYDV-PAV MP may also help transport the viral genome into the nucleus as the MP is present in the cytoplasm and the nucleus [51]. After entry into the cytoplasm, protein synthesis is initiated [52] and enhances replication and transcription efficiency; viruses use the strategy of compartmentalization in specific intracellular components [53]. Its replication is almost totally restricted within the plant phloem tissue [54, 55], i.e., phloem parenchyma cells, companion cells, and sieve tubes. The restricted site of infection in phloem tissue is an essential feature of the *Luteoviridae* [56].

The systemic spread is suspected to be associated with vascular transport of virions due to the discovery of BYDV particles in vasculature samples [57, 58]. The critical role of MP was emphasized by the association between the long-distance movement of some viruses and viral gene expression. For example, geminiviruses coded two proteins responsible for long-distance transportation and viral DNA with a single-stranded genome in and out of the nuclei [59, 60]. However, studies of the function of putative luteoviral MPs remain limited [61]. A 17 kDa protein encoded by ORF 4 is required for BYDV-PAV to spread systemically in plants [62, 63]. The replication of the plant virus genome occurs in host cells [64]. The genome of viruses must be transported into the nucleus by mechanisms requiring viral MP [65].

Three stages of infection were proposed by [57, 33]. At the first stage, densely staining material appeared in plasmodesmata, and an amorphous substance and viral RNA containing filaments appeared in the host cytoplasm, Figures 5 (I)-(M) in chapter 3. In the second stage, filaments became visible in the nuclear pores. During this stage, the nuclear outline became distorted and massive clumping of heterochromatin occurs, Figures (5) and (6) in chapter 1. In the nucleus, viral particles were seen at the last stage after the disintegration of the nuclear membrane. The virus could only infect phloem parenchyma, sieve elements, and companion cells, while it could not be seen in the mestome sheath or the xylem.

The characterization of VLPs was defined by their diameter, their circular outlines, and high electron opacity. Viral particles were detected within areas containing filamentous material Figures (2) and (3) in chapter 4.

Our study concludes that Nanoscience leads to developing a range of inexpensive nanotech applications for enhanced plant growth. The included data proved an efficient means to control virus infection in a fashion way to reduce collateral damage. AuNPs have a dual positive effect on controlling the plant viral disease and enhance strong plant growth performance.

Transmission electron microscope	**TEM**
Gold nanoparticles	**AuNPs**
King Saud University	**KSU**
Barley Yellow Dwarf Virus	**BYDV-PAV**
Reactive Oxygen Species	**ROS**
Endoplasmic reticulum	**ER**
Silver	**Ag**
Gold	**Au**
Iron	**Fe**
Cobalt	**Co**
Copper oxide	**CuO**
Hydroxide	**OH**

NanoMeter	**nm**
Silver nanoparticles	**AgNPs**
Engineered nanomaterial	**ENM**
Metal nanoparticles	**MeNPs**
Open Reading Frame	**ORF3**
Coat protein	**CP**
kilodalton	**kDa**
Ribonucleic Acid	**RNA**
Potato leafroll Polerovirus	**PLRV**
Movement protein	**MP**
Tobacco mosaic virus	**TMV**
Deoxyribonucleic Acid	**DNA**

Author details

Noorah Abdulaziz Othman Alkubaisi[1*] and Nagwa Mohammed Amin Aref[2]

1 Department of Botany and Microbiology, College of Science, King Saud University, Riyadh, Kingdom of Saudi Arabia

2 Department of Microbiology, College of Agriculture, Ain Shams University, Cairo, Egypt

*Address all correspondence to: nalkubaisi@ksu.edu.sa

References

[1] N. Alkubaisi, N. Aref, A. Hendi, Method of inhibiting plant virus using gold nanoparticles, US Patents US9198434B1 1 (2015).

[2] NA. Alkubaisi, N.M. Aref, Dispersed gold nanoparticles potentially ruin gold barley yellow dwarf virus and eliminate virus infectivity hazards, Applied Nanoscience 7(1-2) (2017) 31-40.

[3] W. Falco, E. Botero, E. Falcão, E. Santiago, V. Bagnato, A. Caires, In vivo observation of chlorophyll fluorescence quenching induced by gold nanoparticles, Journal of photochemistry and Photobiology A: Chemistry 225(1) (2011) 65-71.

[4] W. Rong, X. Wang, X. Wang, S. Massart, Z. Zhang, Molecular and ultrastructural mechanisms are underlying yellow dwarf symptom formation in wheat after barley infection yellow dwarf virus, International journal of molecular sciences 19(4) (2018) 1187.

[5] P.H. Nass, L.L. Domier, B.P. Jakstys, C.J. D'Arcy, In situ localization of barley yellow dwarf virus-PAV 17-kDa protein and nucleic acids in oats, Phytopathology 88(10) (1998) 1031-1039.

[6] N. Hardwick, D. Jones, J. Slough, Factors affecting diseases of winter wheat in England and Wales, 1989-98, Plant Pathology 50(4) (2001) 453-462.

[7] N. Pal, J.S. Moon, J. Sandhu, L.L. Domier, C.J. D'Arcy, Production of Barley yellow dwarf virus antisera by DNA immunization, Canadian Journal of Plant Pathology 22(4) (2000) 410-415.

[8] A.M. French, California plant disease host index, California Department of Food and Agriculture, Division of Plant Industry1989.

[9] M. Vargas-Hernandez, I. Macias-Bobadilla, R.G. Guevara-Gonzalez, E. Rico-Garcia, R.V. Ocampo-Velazquez, L. Avila-Juarez, I. Torres-Pacheco, Nanoparticles as Potential Antivirals in Agriculture, Agriculture 10(10) (2020) 444.

[10] T. Karuppanapandian, J.-C. Moon, C. Kim, K. Manoharan, W. Kim, Reactive oxygen species in plants: their generation, signal transduction, and scavenging mechanisms, Australian Journal of Crop Science 5(6) (2011) 709.

[11] E. Pradedova, O. Isheeva, R. Salyaev, Classification of the antioxidant defense system as the ground for the reasonable organization of experimental studies of the oxidative stress in plants, Russian Journal of plant physiology 58(2) (2011) 210-217.

[12] Y. Zhai, X. Liu, H. Chen, B. Xu, L. Zhu, C. Li, G. Zeng, Source identification and potential ecological risk assessment of heavy metals PM2. 5 from Changsha, Science of the Total Environment 493 (2014) 109-115.

[13] M.H. Siddiqui, M.H. Al-Whaibi, M. Firoz, M.Y. Al-Khaishany, Role of nanoparticles in plants, Nanotechnology and Plant Sciences, Springer2015, pp. 19-35.

[14] K.-J. Dietz, S. Herth, Plant nanotoxicology, Trends in plant science 16(11) (2011) 582-589.

[15] Z. Guo, H. Xu, S. Su, J. Cai, S. Dang, S. Xiang, G. Qian, H. Zhang, M. O'Keeffe, B. Chen, A robust near-infrared luminescent ytterbium metal-organic framework for sensing of small molecules, Chemical Communications 47(19) (2011) 5551-5553.

[16] G.X. Shen, Oxidative stress and diabetic cardiovascular disorders: roles of mitochondria and NADPH oxidase,

Canadian journal of physiology and pharmacology 88(3) (2010) 241-248.

[17] D.H. Atha, H. Wang, E.J. Petersen, D. Cleveland, R.D. Holbrook, P. Jaruga, M. Dizdaroglu, B. Xing, B.C. Nelson, Copper oxide nanoparticle mediated DNA damage in terrestrial plant models, Environmental science & technology 46(3) (2012) 1819-1827.

[18] J. Koelmel, T. Leland, H. Wang, D. Amarasiriwardena, B. Xing, Investigation of gold nanoparticles uptake and their tissue level distribution in rice plants by laser ablation-inductively coupled-mass spectrometry, Environmental pollution 174 (2013) 222-228.

[19] S. Tatur, M. Maccarini, R. Barker, A. Nelson, G. Fragneto, Effect of functionalized gold nanoparticles on floating lipid bilayers, Langmuir 29(22) (2013) 6606-6614.

[20] T. Silva, L.R. Pokhrel, B. Dubey, T.M. Tolaymat, K.J. Maier, X. Liu, Particle size, surface charge and concentration-dependent ecotoxicity of three organo-coated silver nanoparticles: comparison between general linear model-predicted and observed toxicity, Science of the Total Environment 468 (2014) 968-976.

[21] J.E. Miller, Effects on photosynthesis, carbon allocation, and plant growth associated with air pollutant stress, Assessment of crop loss from air pollutants, Springer1988, pp. 287-314.

[22] KVK. Mohan, B. Ghebrehiwet, C.D. Atreya, The N-terminal conserved domain of rubella virus capsid interacts with the C-terminal region of cellular p32 and overexpression of p32 enhances the viral infectivity, Virus research 85(2) (2002) 151-161.

[23] C. Chay, U. Gunasinge, S. Dinesh-Kumar, W. Miller, S.M. GRAY, Aphid transmission and systemic plant infection determinants of barley yellow dwarf luteovirus-PAV are contained in the coat protein readthrough domain and 17-kDa protein, respectively, Virology 219(1) (1996) 57-65.

[24] R. Di, S. Dinesh-Kumar, W. Miller, Translational frameshifting by barley yellow dwarf virus RNA (PAV serotype) in Escherichia coli and in eukaryotic cell-free extracts, Molecular plant-microbe interactions: MPMI 6(4) (1993) 444.

[25] E. Tacke, J. Schmitz, D. Prüfer, W. Rohde, Mutational analysis of the nucleic acid-binding 17 kDa phosphoprotein of potato leafroll luteovirus identifies an amphipathic α-helix as the domain for protein/protein interactions, Virology 197(1) (1993) 274-282.

[26] E. Tacke, D. Prüfer, J. Schmitz, W. Rohde, The potato leafroll luteovirus 17K protein is a single-stranded nucleic acid-binding protein, Journal of General Virology 72(8) (1991) 2035-2038.

[27] D. Beebe, W. Russin, Plasmodesmata in the phloem-loading pathway, Plasmodesmata, Springer1999, pp. 261-293.

[28] S.-L. Cheng, L.L. Domier, C.J. D'Arcy, Detection of the readthrough protein of barley yellow dwarf virus, Virology 202(2) (1994) 1003-1006.

[29] S. Dinesh-Kumar, V. Brault, W. Allen, Precise mapping and in vitro translation of a trifunctional subgenomic RNA of barley yellow dwarf virus, Virology 187(2) (1992) 711-722.

[30] S. Filichkin, R. Lister, P. McGrath, M. Young, In vivo expression and mutational analysis of the barley yellow dwarf virus readthrough gene, Virology 205(1) (1994) 290-299.

[31] P. Larkin, M. Young, W. Gerlach, P. Waterhouse, The Yd2 resistance to barley yellow dwarf virus is effective in barley plants but not in their leaf protoplasts, Annals of applied biology 118(1) (1991) 115-125.

[32] B. Mohan, S. Dinesh-Kumar, W.A. Miller, GENES AND CZS-ACTING SEQUENCES INVOLVED IN REPLICATION OF BARLEY YELLOW DWARF VIRUS-PAV RNA, Replication of barley yellow dwarf luteovirus-PAV RNA (1995) 58.

[33] S. Choudhury, H. Hu, H. Meinke, S. Shabala, G. Westmore, P. Larkin, M. Zhou, Barley yellow dwarf viruses: infection mechanisms and breeding strategies, Euphytica 213(8) (2017) 168.

[34] E. Waigmann, S. Ueki, K. Trutnyeva, V. Citovsky, The ins and outs of nondestructive cell-to-cell and systemic movement of plant viruses, Critical Reviews in Plant Sciences 23(3) (2004) 195-250.

[35] S. Ghoshroy, R. Lartey, J. Sheng, V. Citovsky, Transport of proteins and nucleic acids through plasmodesmata, Annual review of plant biology 48(1) (1997) 27-50.

[36] C.M. Deom, M.J. Oliver, R.N. Beachy, The 30-kilodalton gene product of tobacco mosaic virus potentiates virus movement, Science 237(4813) (1987) 389-394.

[37] T. Meshi, Y. Watanabe, T. Saito, A. Sugimoto, T. Maeda, Y. Okada, Function of the 30 kd protein of tobacco mosaic virus: involvement in the cell-to-cell movement and dispensability for replication, The EMBO Journal 6(9) (1987) 2557-2563.

[38] W. Jin, X. Gu, S. Li, P. Huang, N. Xu, J. Shi, Experimental and simulation study on a catalyst packed tubular dense membrane reactor for partial oxidation of methane to syngas, Chemical Engineering Science 55(14) (2000) 2617-2625.

[39] V. Haywood, F. Kragler, W.J. Lucas, Plasmodesmata: pathways for protein and ribonucleoprotein signaling, The Plant Cell 14(suppl 1) (2002) S303-S325.

[40] V. Citovsky, D. Knorr, G. Schuster, P. Zambryski, The P30 movement protein of tobacco mosaic virus is a single-strand nucleic acid-binding protein, Cell 60(4) (1990) 637-647.

[41] V. Citovsky, M.L. Wong, A.L. Shaw, B. Prasad, P. Zambryski, Visualization and characterization of tobacco mosaic virus movement protein binding to single-stranded nucleic acids, The Plant Cell 4(4) (1992) 397-411.

[42] O. Kiselyova, I. Yaminsky, E. Karger, O.Y. Frolova, Y.L. Dorokhov, J. Atabekov, Visualization by atomic force microscopy of tobacco mosaic virus movement protein–RNA complexes formed in vitro, Journal of General Virology 82(6) (2001) 1503-1508.

[43] M. Heinlein, B.L. Epel, H.S. Padgett, R.N. Beachy, Interaction of tobamovirus movement proteins with the plant cytoskeleton, Science 270(5244) (1995) 1983-1985.

[44] B.G. McLean, J. Zupan, P.C. Zambryski, Tobacco mosaic virus movement protein associates with the cytoskeleton in tobacco cells, The Plant Cell 7(12) (1995) 2101-2114.

[45] E. Waigmann, A. Turner, J. Peart, K. Roberts, P. Zambryski, Ultrastructural analysis of leaf trichome plasmodesmata reveals major differences from mesophyll plasmodesmata, Planta 203(1) (1997) 75-84.

[46] S. Malyshenko, O. Kondakova, J.V. Nazarova, I. Kaplan, M. Taliansky, J. Atabekov, Reduction of tobacco mosaic virus accumulation in transgenic plants producing non-functional viral

transport proteins, Journal of general virology 74(6) (1993) 1149-1156.

[47] J. Schmitz, C. Stussi-Garaud, E. Tacke, D. Prüfer, W. Rohde, O. Rohfritsch, In SituLocalization of the putative movement protein (pr17) from potato Leafroll Luteovirus (PLRV) in infected and transgenic potato plants, Virology 235(2) (1997) 311-322.

[48] T. Xia, M. Kovochich, M. Liong, L. Madler, B. Gilbert, H. Shi, J.I. Yeh, J.I. Zink, A.E. Nel, Comparison of the mechanism of toxicity of zinc oxide and cerium oxide nanoparticles based on dissolution and oxidative stress properties, ACS Nano 2(10) (2008) 2121-2134.

[49] J. Han, Y. Lee, K.-H. Yeom, Y.-K. Kim, H. Jin, V.N. Kim, The Drosha-DGCR8 complex in primary microRNA processing, Genes & development 18(24) (2004) 3016-3027.

[50] M.C.S. Lee, E.A. Miller, J. Goldberg, L. Orci, R. Schekman, BI-DIRECTIONAL PROTEIN TRANSPORT BETWEEN THE ER AND GOLGI, Annual Review of Cell and Developmental Biology 20(1) (2004) 87-123.

[51] P.H. Nass, L.L. Domier, B.P. Jakstys, C.J. D'Arcy, In Situ Localization of Barley Yellow Dwarf Virus-PAV 17-kDa Protein and Nucleic Acids in Oats, Phytopathology® 88(10) (1998) 1031-1039.

[52] J.A.d. Boon, P. Ahlquist, Organelle-Like Membrane Compartmentalization of Positive-Strand RNA Virus Replication Factories, Annual Review of Microbiology 64(1) (2010) 241-256.

[53] R.R. Novoa, G. Calderita, R. Arranz, J. Fontana, H. Granzow, C. Risco, Virus factories: associations of cell organelles for viral replication and morphogenesis, Biology of the Cell 97(2) (2005) 147-172.

[54] M. Irwin, J. Thresh, Epidemiology of barley yellow dwarf: a study in ecological complexity, Annual review of phytopathology 28(1) (1990) 393-424.

[55] G. Delogu, L. Cattivelli, M. Snidaro, A. Stanca, The Yd2 gene and enhanced resistance to barley yellow dwarf virus (BYDV) in winter barley, Plant breeding 114(5) (1995) 417-420.

[56] J. Kehr, A. Buhtz, Long-distance transport and RNA movement through the phloem, Journal of experimental botany 59(1) (2008) 85-92.

[57] C. Gill, J. Chong, Development of the infection in oat leaves inoculated with barley yellow dwarf virus, Virology 66(2) (1975) 440-453.

[58] S.G. Jensen, Systemic movement of barley yellow dwarf virus in small grains, Phytopathology 63(7) (1973) 854-856.

[59] A.O. Noueiry, W.J. Lucas, R.L. Gilbertson, Two proteins of a plant DNA virus coordinate nuclear and plasmodesmal transport, Cell 76(5) (1994) 925-932.

[60] E. Pascal, A.A. Sanderfoot, B.M. Ward, R. Medville, R. Turgeon, S.G. Lazarowitz, The geminivirus BR1 movement protein binds single-stranded DNA and localizes to the cell nucleus, The Plant Cell 6(7) (1994) 995-1006.

[61] Z. Xia, Y. Wang, Z. Du, J. Li, R.Y. Zhao, D. Wang, A potential nuclear envelope-targeting domain and an arginine-rich RNA binding element identified in the putative movement protein of the GAV strain of Barley yellow dwarf virus, Functional Plant Biology 35(1) (2008) 40-50.

[62] C.A. Chay, D.M. Smith, R. Vaughan, S.M. Gray, Diversity among isolates within the PAV serotype of barley yellow dwarf virus, Phytopathology 86(4) (1996) 370-377.

[63] G. Koev, B. Mohan, S. Dinesh-Kumar, K.A. Torbert, D.A. Somers, W.A. Miller, Extreme reduction of disease in oats transformed with the 5′ half of the barley yellow dwarf virus-PAV genome, Phytopathology 88(10) (1998) 1013-1019.

[64] L. Hanley-Bowdoin, S.B. Settlage, B.M. Orozco, S. Nagar, D. Robertson, Geminiviruses: models for plant DNA replication, transcription, cell cycle regulation, Critical Reviews in Plant Sciences 18(1) (1999) 71-106.

[65] H.B. Scholthof, Plant virus transport: motions of functional equivalence, Trends in plant science 10(8) (2005) 376-382

Chapter 2

Nucleus

Noorah Abdulaziz Othman Alkubaisi
and Nagwa Mohammed Amin Aref

Abstract

In our application of AuNPs on the leaf surface, we were pushing the Barley Yellow Dwarf Virus (BYDV-PAV) source and Gold nanoparticles (AuNPs) into the plant cell system up on the events of systemic plant defense response. In the infected host cell, the viral coat protein is the first obvious in the cytoplasm. When nanoparticles are applied on leaf surfaces, a large surface area relative to their volume happens. AuNPs solutions are more active and dispersed ooplasm. The correlation between Zeta potential value and Zeta sizer is inverse proportion. Filaments are visible in the nucleopores, the nuclear outline is distorted, and massive clumping of heterochromatin begins as declared. It was mostly found in or around regions of ribosome-associated filaments. Our present study combines TEM and nucleus content in the presence of AuNPS to explore the level of repair mechanism illustrating in TEM micrographs, showing Polyploidy nucleus and segregated chromatin. Multi membranous structure, imaging the AuNPs inside and around the nucleus and Pseudo crystal array is enveloped in an endoplasmic reticulum cisterna (ER).

Keywords: Nucleus, Barley Yellow Dwarf Virus (BYDV-PAV), Gold nanoparticles (AuNPs), Mechanically inoculation, Inclusions body, Segregated chromatin, Polyploidy nucleus, Spindle shape, Endosomes, Multi membranous structure, Rhombic crystal array

1. Introduction

The study succeeded in inducing BYDV-PAV infection mechanically [1] on the surface. Negative staining preparation for Electron Microscopy is used for staining virus particles and the morphological and cytological side of healthy (**Figure 1**) and treated leaves [2]. It is proposed that coat protein is expressed by cytoplasmic ribosomes from viral RNA coat protein found in the nucleus during later stages of infection probably diffused into the nucleoplasm after disruption of the nuclear membrane as evident in **Figure 2**; virus particles were then also numerous in the pockets of the nucleoplasm as shown in **Figures 2(B)** and **3(A)** and **(C)**. The studied virus revealed in ultrastructure preparations characteristic performance in the infected tissue of barely plant, crystalline array, numerous slender filamentous shape inclusions, as in **Figure 5(A)–(C)**, proteinous content, amorphous material, some cytoplasmic components which take irregular shape and inclusion bodies as appeared in **Figure 6(A)** and **(B)**. These Cytopathic effects of BYDV-PAV resemble the same as [3], who studied the ultrastructure of infected cells and confirm the restriction of BYDV-PAV to phloem parenchyma, companion cells, and sieve elements of leaves. As long as the value of the surface potential of AuNPs in mv is high,

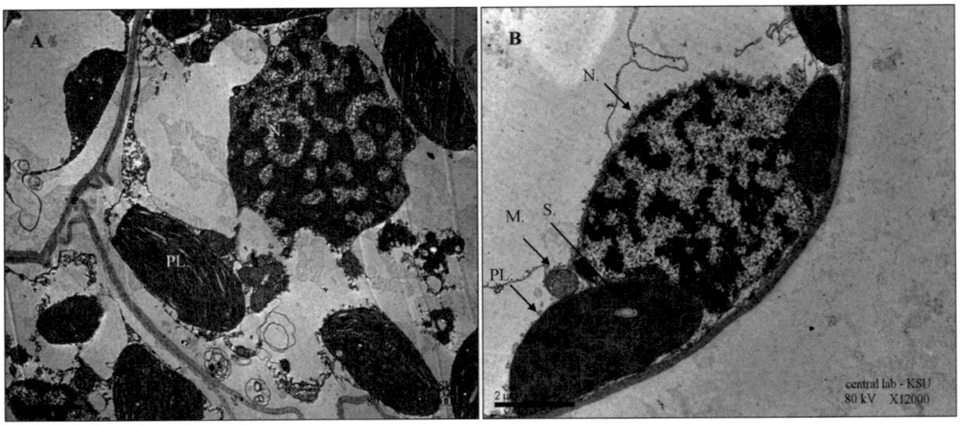

Figure 1.
High-resolution transmission electron microscopy imaging a general view of Hordeum vulgare; *barley healthy cells from leaves. (A and B) The left and right nucleus view of barley cell incubated with a single nucleus with different magnification (10000, 12000 kV). Scale bar 2 µm. The solid arrows indicate nucleus (N), chloroplast (PL), starch granules (S). Scale bar 2 µm.*

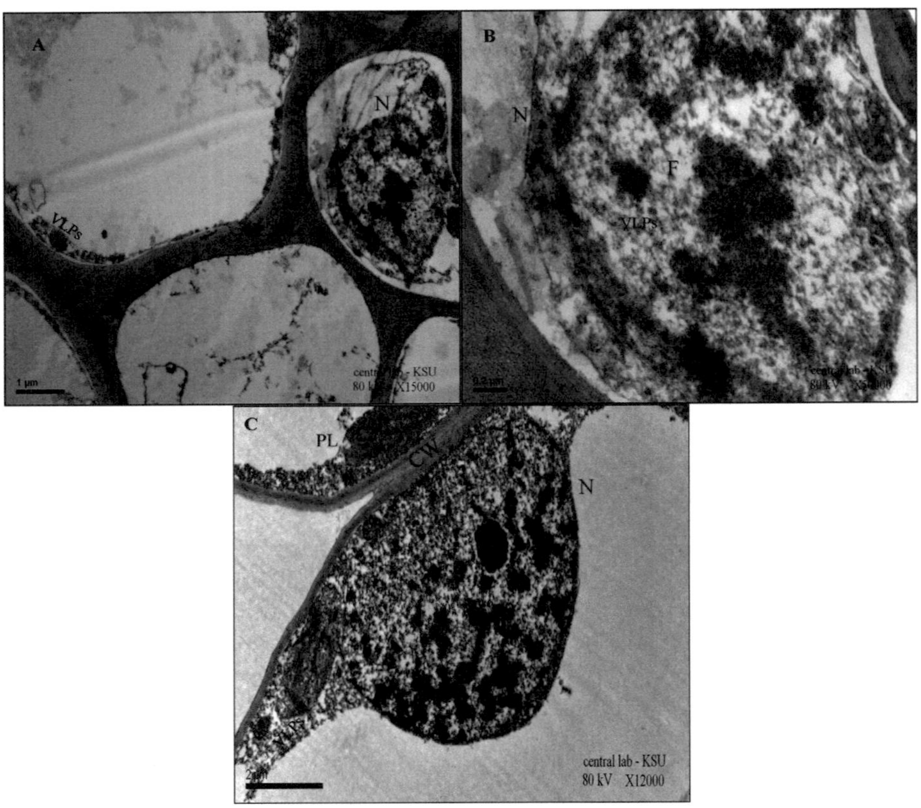

Figure 2.
Transmission electron micrographs showing deformation of the nucleus from infected leaves by barley yellow dwarf virus (BYDV). (A) A general view inside the cells showed the large nucleus and nucleolus with segregated distinct chromatin. Scale bar 1 µm. (B) a higher magnification (50000, 12000 kV) of the previous picture A scale bar 0.2 µm. (C) Showed segregated chromatin of the nucleus beside the cell wall. Scale bar 2 µm.

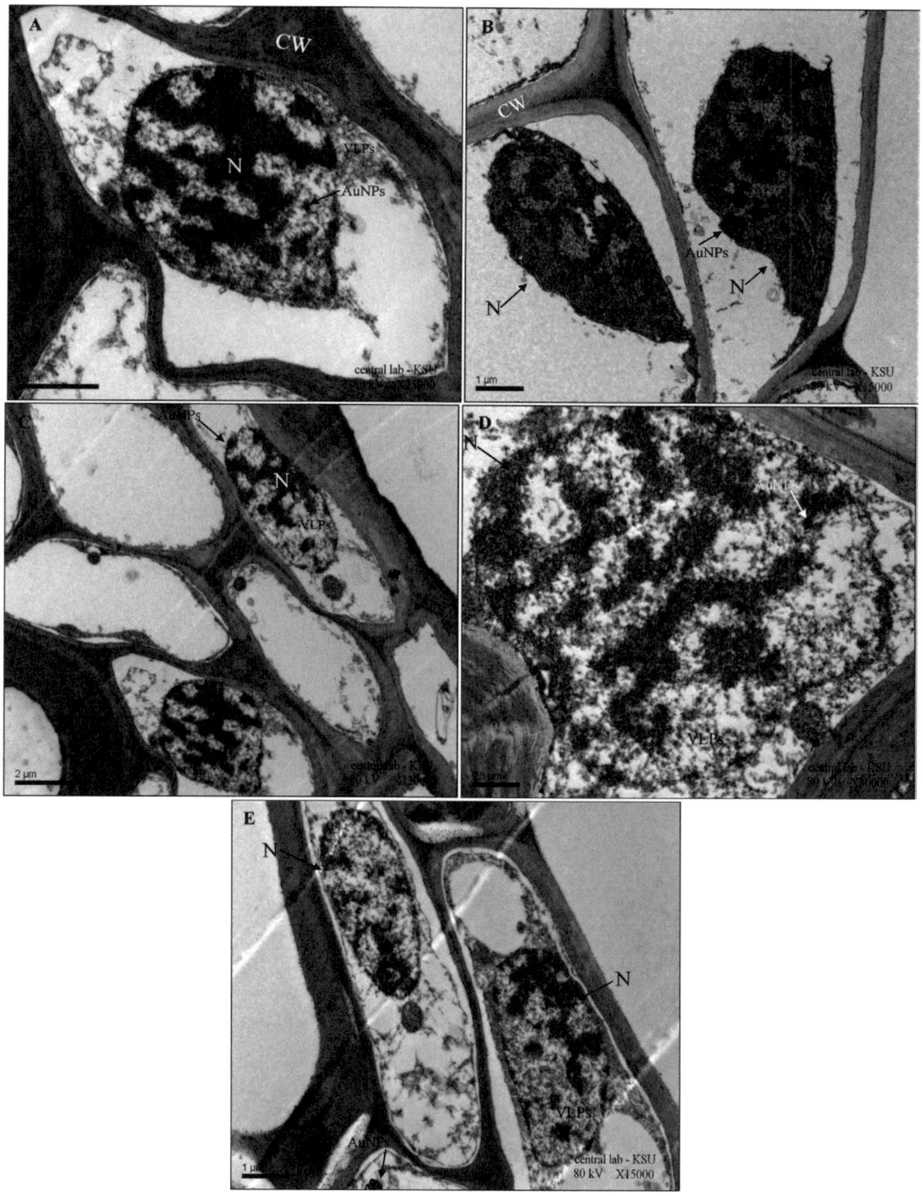

Figure 3.
Polyploidy nucleus and segregated chromatin (showing the lighter staining and partially eroded heterochromatin). (A) Different shaped of polyploidy nucleus and segregated chromatin infected with the virus (VLPs) and pretreated with gold nanoparticles (AuNPs). Scale bar 1 µm. (B) Spindle shape (arrows). Scale bar 1 µm. (C) Perpendicular shape (arrow). Scale bar 2 µm. (D) Abnormal shape of the nucleus, spindle shape (arrow). Scale bar 1 µm. (E) Abnormal shape of the nucleus; ovule shape (arrow) with more isolated chromatin. Scale bar 0.5 µm.

the solutions could be colloids/AuNPs.sol. The tiny size of nanoparticles means they exhibit enhanced or different properties compared with the bulk material, **Figure 4(A)–(D)**. Nanoparticles also enter through the stomata openings or the bases of trichomes and then translocated to various tissues mentioned.

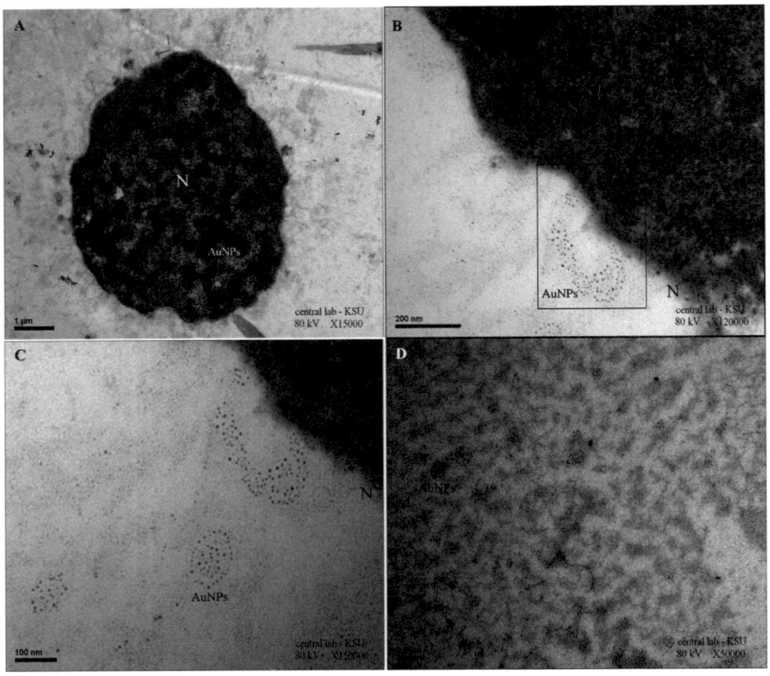

Figure 4.
Transmission electron microscopy imaging the AuNPs inside the nucleus. *(A and B) Micrograph of the polyploidy rounded nucleus filled with AuNPs. Scale bar 1 µm. (B and C) Transmission electron microscopy imaging the AuNPs around the nucleus. A micrograph of the organization of the AuNPs inside the cells, AuNPs gathering irregularly around the nucleus. Scale bar 200 nm, 100 nm. (D) Highly existence of AuNPs around the nucleus. Scale bar 0.2 µm.*

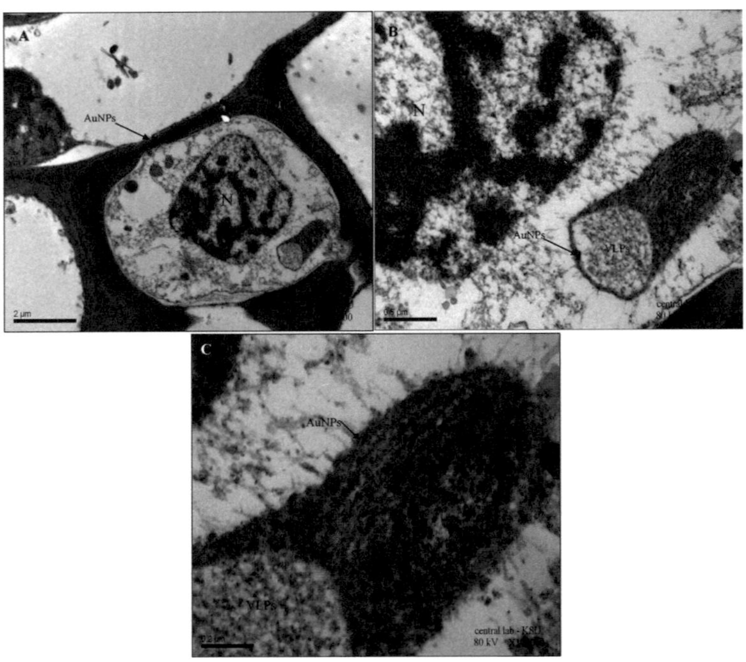

Figure 5.
Micrographs of pseudo crystal array which is enveloped in an endoplasmic reticulum cisterna (ER). *(A) Abnormal shape of the nucleus with a perpendicular line with more isolated chromatin. Pseudo crystal array near the nucleus. Scale bar 2 µm. (B) Higher magnification of the previous picture A. Scale bar 0.5 µm. (C) Micrograph of higher magnification for rhombic crystal array contains a lot of AuNPs and VLPs inside the crystal. Scale bar 0.2 µm.*

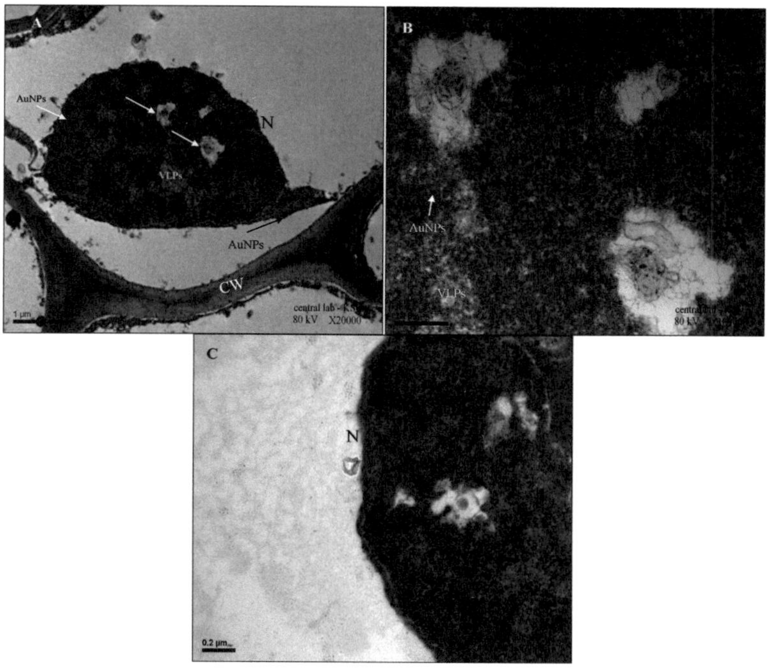

Figure 6.
Multi membranous structure. *(A) Micrograph inside the spindle nucleus has a multi membranous structure contains bodies like endosomes (arrows). Scale bar 1 μm. (B) a higher magnification (20000, 100000 kV) of these bodies. Scale bar 0.2 μm. (C) a higher magnification (6000 kV) of the irregular distribution of the chromatin with empty spaces inside the nucleus having some vacuoles. Scale bar 0.2 μm.*

2. Hyper polyploidy nucleus

Three phases of infection were defined based on alternation in the cytoplasm (early phase), nucleus (intermediate), **Figure 2(B)** and **(C)**, and both (late). The significant changes during infection with BYDV-PAV begin with the appearances of densely staining material in plasmodesmata, amorphous substance, and filaments, vesicles in the cytoplasm [3]. They suggested that the cytoplasm is the site of coat protein expression and viral assembly. At the end of the early stage, filaments are visible in the nucleopores as shown in **Figure 6(B),** During the intermediate, the nuclear outline is distorted, and massive clumping of heterochromatin begins as declared in **Figure 3**. In the present study, it was noticed in massive hyper polyploidy nucleus conjugated with the application of AuNPs in many treatments, **Figure 3(A)–(D)**.

Abbreviations

AuNPs	Gold nanoparticles
BYDV-PAV	Barley Yellow Dwarf Virus
RNA	RiboNucleic Acid
N	Nucleus
S	Starch
PL	Plastid
OG	Osmophilic Globule
MLB	Myelin like bodies

G	Grana
VLPs	Virus-like particles
CW	Cell Wall
SG	Starch granules.
IRR.S	Irregular starch granules

Author details

Noorah Abdulaziz Othman Alkubaisi* and Nagwa Mohammed Amin Aref

1 Department of Botany and Microbiology, College of Science, Female Scientific and Medical Colleges, King Saud University, Riyadh, Kingdom of Saudi Arabia

2 Faculty of Agriculture, Department of Microbiology, Ain Shams University, Muhafazat al Qahirah, Cairo, Egypt

*Address all correspondence to: nalkubaisi@ksu.edu.sa

References

[1] N.A. Alkubaisi, N.M. Aref, Dispersed gold nanoparticles potentially ruin gold barley yellow dwarf virus and eliminate virus infectivity hazards, Applied Nanoscience 7(1-2) (2017) 31-40.

[2] C. Gill, J. Chong, Development of the infection in oat leaves inoculated with barley yellow dwarf virus, Virology 66(2) (1975) 440-453.

[3] V. Mishra, R.K. Mishra, A. Dikshit, A.C. Pandey, Interactions of nanoparticles with plants: an emerging prospective in the agriculture industry, Emerging technologies and management of crop stress tolerance, Elsevier2014, pp. 159-180.

Chapter 3

Chloroplast and Mitochondria

Noorah Abdulaziz Othman Alkubaisi
and Nagwa Mohammed Amin Aref

Abstract

Photosynthesis is a crucial process for plants on earth that changes light energy to chemical energy. Virus infection can cause dramatic photosynthesis changes: respiration and the translocation of carbohydrates and other substances around the host plant. Chlorosis in virus-infected leaves like Barley Yellow Dwarf Virus (BYDV- PAV).infection can result from damage to chloroplasts resulting from inhibition of photosynthetic activity. Our present study combines TEM and chlorophyll-level content in the presence of Gold nanoparticles (AuNPS) to explore the repair mechanism for the yellowing leaf symptom development caused by infection with BYDV- PAV by illustrating TEM micrographs; showing fragmentized grana, deformation of the myelin like bodies (MLB), many vesicles; osmiophilic lipid granules/plastoglobulus, starch body, and plasmolysis in the chloroplast, distribution of AuNPs & VLPs near and inside the chloroplast. Mitochondria, Double-membrane-bound organelle, Distorted mitochondrion, Amorphous inclusion bodies.

Keywords: Chloroplast, Barley Yellow Dwarf Virus (BYDV-PAV), Gold nanoparticles (AuNPs), Starch granules (S), osmiophilic lipid granules/plastoglobulus (OG), Myelin like bodies (MLB), Plasmolysis, Portentous content, Fragmentized grana. Mitochondria, Double-membrane-bound organelle, Distorted mitochondrion, Amorphous inclusion bodies

1. Introduction

BYDV-PAV causes cytological alterations, physiological and biochemical, including the restriction of photosynthate transportation, phloem degeneration, and creating a nutritionally [1], and the formation of specific cytological inclusions [2]. In the susceptible wheat and after BYDV-GAV infection, drop expressions of chlorophyll biosynthesis and chloroplast was noticed due to related genes and altered expression of the ABA mentioned above. Chloroplast in the healthy cells has normal densities of thylakoid in chloroplast (**Figure 1(A)**–**(D)**). The lower chlorophyll content, fragmentized chloroplasts, ROS accumulation, and slower growth could be engaged to ET signaling and ROS related genes, consequently resulting in leaf yellowing and plant dwarfing (**Figure 2**).

Studies by several laboratories of TMV-infected tobacco plants suggest that the uptake of viral CP causes chlorosis in these plants by thylakoid. R. Beachy's lab observed that the CP aggregates in the thylakoid membrane. Resulting the intense free radical damage to the organelle refer to pull apart photosystem II (the water-splitting reaction of photosynthesis). In vitro experiments, M. Zaitlin's laboratory

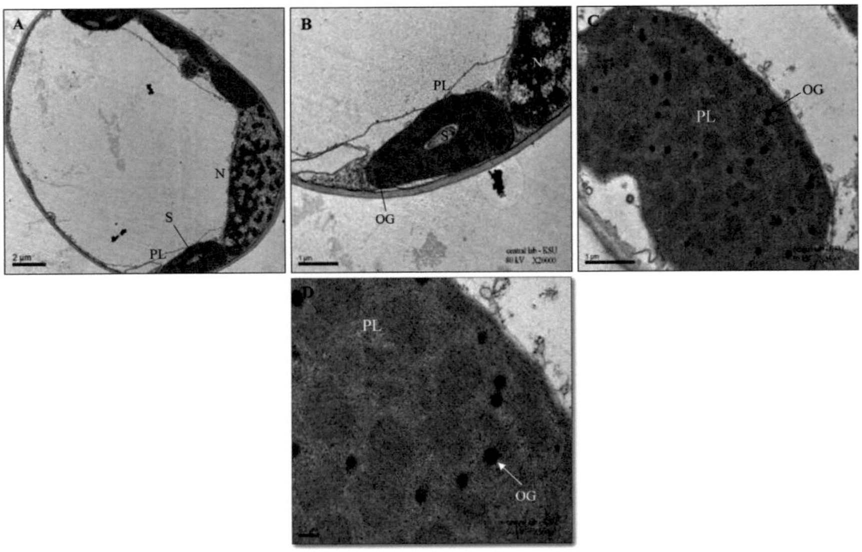

Figure 1.
Electron micrographs showing a general view inside the healthy cells of barley leaves with densities of thylakoids in chloroplasts. **(A)** *The cells show the nucleus and chloroplast with one starch granules, with low magnification, and higher magnification in* **(B)***, Scale bars 2 µm and 1 µm.* **(C)** *The outline area shows the inside details of the grana, Scale bar 1 µm.* **(D)** *The cell shows the plastid contains the osmiophilic lipid granules/plastoglobulus and higher magnification of the grana structure, Scale bar 0.2 µm.*

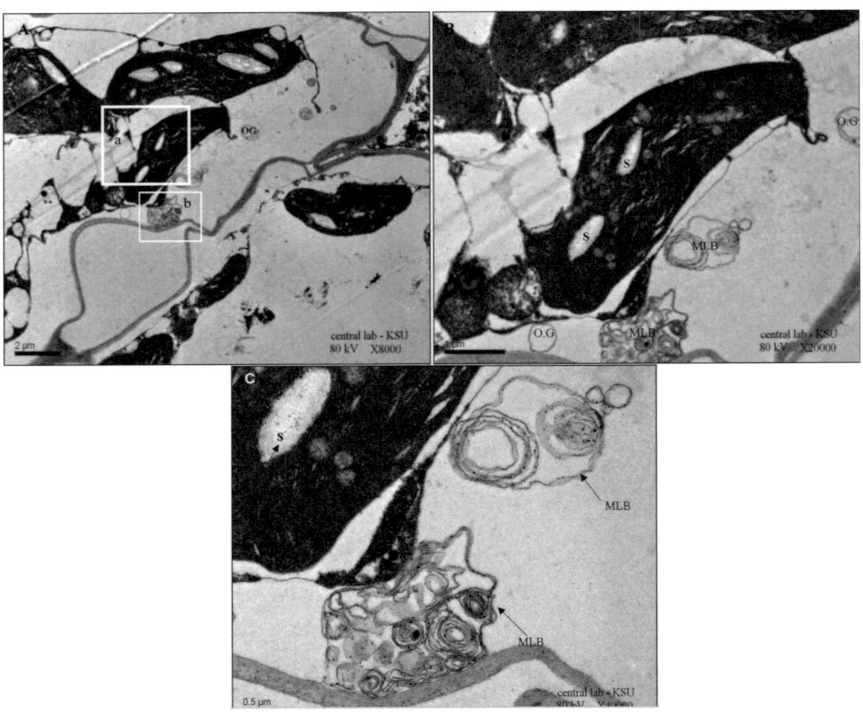

Figure 2.
Electron micrographs showing fragmentized grana. **(A)** *A general view inside the cell infected with BYDV- PAV at lower magnification [outlined area is shown at higher magnification in A]. Scale bar 2 µm.* **(B)** *Accumulation of starch granules and many osmophilic globules (OGs) is observed in the chloroplast surrounded by bundle sheath cells integrated into the infected cell, Scale bar 1 µm. Electron micrograph showing the deformation of the myelin like bodies* **(C)** *Numerous intracellular inclusion bodies myelin like bodies (MLB), (arrows), Scale bar 0.5 µm.*

noticed that uptake of the CP did not accomplish the host's normal pathways for protein import using isolated chloroplasts. Instead, a feature of the CP, when it was assembled into multimeric disks (an intermediate structure in TMV virion assembly), let it be taken up randomly into the chloroplasts and the thylakoid membrane eventually. The single amino acid in the CP determines its uptake efficiency into chloroplasts, and K. Lehto has detected the relative severity of chlorosis caused by different TMV strains. Changes in photosynthesis, respiration, and carbohydrate translocation generally cause localized aggregation and depletion of starch, the primary stock carbohydrate of most plants [3]. Common cytopathological effects of BYDV-PAV infection are restricted to the phloem of host plants cleared [4], where they are seen via EM in the cytoplasm, nuclei, and vacuoles of infected sieve elements, companion and parenchyma cells. Vesicles containing filaments and inclusions containing virus particles the infection and subsequent death of phloem cells inhibits translocation as **Figure 2(A)–(C)**, slows plant growth, and induces loss of chlorophyll typical symptoms. Plants convert only 2–4% of the available energy in radiation into new plant growth [5].

2. Chemical energy production of nanoparticles

Metal nanoparticles can induce the efficiency of chemical energy production in photosynthetic systems, **Figure 5(C)**. Nano-anatase TiO2 has a photocatalyzed characteristic, improves the light absorbance, the transformation from light energy to electrical, chemical energy, and induces carbon dioxide assimilation. TiO2NPs protect chloroplast from aging for long time illumination [6, 7]. Nano-anatase TiO2 enhances the photosynthetic carbon assimilation with potentially activate Rubisco (a complex of Rubisco [8], which promotes Rubisco carboxylation, thereby increasing the growth of plants [8, 9] studied the impact of nano-anatase on the molecular mechanism of carbon reaction and postulated the marker gene's induction for Rubisco activase (RCA) mRNA by nano-anatase. Enhancing protein levels and activities of Rubisco activated the improvement of the Rubisco carboxylation and the high rate of photosynthetic carbon reaction. The exogenous application of TiO2NPs improves the net photosynthetic rate, conductance to water, and transpiration rate in plants [10]. Nano-anatase promoted heartily whole chain electron transport [11], photoreduction activity of photosystem II, O2-evolving, and photophosphorylation activity of chlorophyll under both visible and ultraviolet light. The AuNPs and Ag nanocrystals bind to the chlorophyll in the photosynthetic reaction center [12], forming a novel hybrid system, which may build ten times more excited electrons plasmon resonance fast electron–hole separation. The enhancement mechanisms may help the design of artificial light-harvesting systems. Nanomaterials can generate ROS, affect lipid peroxidation, as reported [13]. The significant biochemical and molecular effect on membrane permeability and fluidity, which due to previous nanomaterials impact making cells more susceptible to osmotic stress and failure to nutrient uptake. A series of metabolic activities such as soil and water which perceived the stress through the growth matrix [14–16] triggered to alleviate the metal stressors [17]. To deal with the situation, in the first step, to prohibit metal entry through the expense of energy, plants modulating their action actively. In the second step, modulating transporters in the plasma membrane prevent further entry of the metal into the cytosol, so that the intracellular build-up of metal ions does not exceed the threshold concentration, **Figure 3(A, B, and D)**. The plant system has developed several well-synchronized systems to Elementary Flux Mode (EFM) to prevent metal ion build-up from the cellular milieu [18].

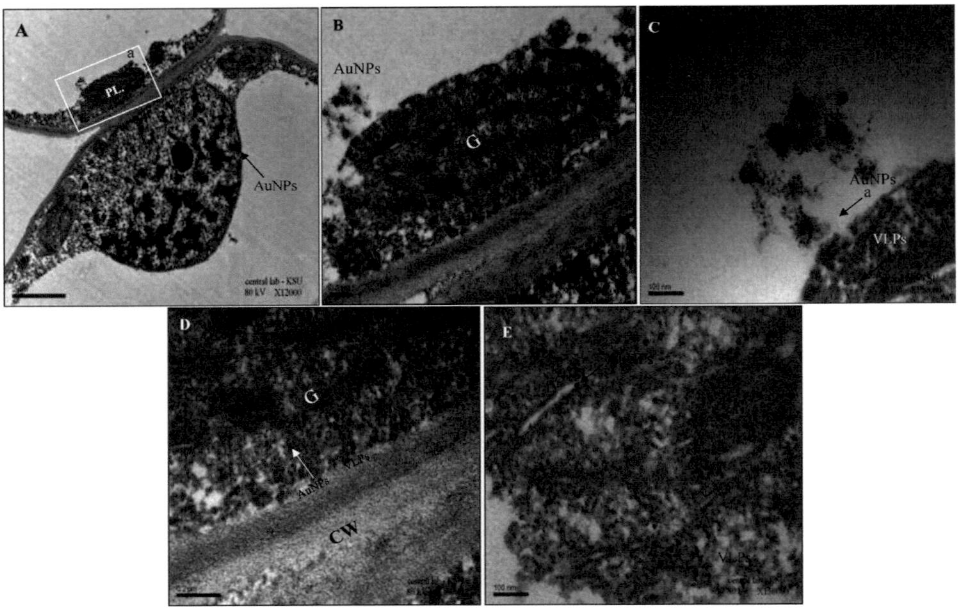

Figure 3.
Micrographs of a different distribution of AuNPs & VLPs near the chloroplast. **(A)** *Micrograph of large nucleus and nucleolus with segregated distinct chromatin beside the cell wall, AuNPs restricted to the chloroplast from both sides, Scale bars 2μm.* **(B)** *The random bulk number of localized AuNPs aggregated intimately near to the chloroplast; Scale bars 0.2 μm.* **(C)** *Packed AuNPs in the cytoplasm adjacent to VLPs, Scale bars 100 nm. Micrographs of a different distribution of AuNPs & VLPs inside the chloroplast.* **(D)** *Dispersed AuNPs into the grana, which appeared to be tiered apart in an unregulated shape, VLPs and AuNPs condensate near the cell wall inside the plastid. Scale bar 0.2 μm.* **(E)** *Micrographs of the precipitated AuNPs and VLPs between the grana structure. Scale bar 100 nm.*

AuNPs significantly increased vegetative growth and seed production in both noncrops (*Arabidopsis thaliana*) [19] and crop (*Brassica juncea*) species [20]. TEM images revealed that chloroplasts in BYDV-GAV-infected Zhong8601 leaf cells were fragmentized [21] as shown in **Figures 3(A)** and **4(B)** and **(D)** as Where thylakoids were not well developed, but starch granules as in **Figures 3(B, E, and F)** and **4(C)** and plastoglobules were not rare in our study. Compared to another strain study for BYDV-GAV [21], mock-inoculated Zhong8601, chlorophyll content was reduced, but the virus and H2O2 contents were markedly higher in BYDV-GAV-infected Zhong860. Reactive oxygen species (ROS)-related genes were transcriptionally regulated in BYDV-GAV infected Zhong8601. These results suggest that the yellow dwarf symptom formation is mainly attributed to reduced chlorophyll content and fragmentized chloroplasts [22]. Phloem damage caused by BYDV limits the transport of photosynthate and restricts long-distance carbohydrate translocation [21]. Carbohydrate accumulation in leaves consecutively inhibits photosynthesis, reduces chlorophyll, and increases respiration [23].

The changes in root system function such as root length or biomass caused by BYDV infection is unknown, although root tips are far from the photosynthate source but suffered from reduced translocation. Potential reductions in root system function of BYDV-infected plants could play a crucial role in grain yield loss because root systems supply shoot organs with fundamental mineral nutrients [24] due to BYDV infection, and a susceptible wheat cultivar showed a 72% reduction in photosynthetic capacity. On the other hand, a moderately tolerant wheat cultivar exhibited only a 60% reduction in photosynthesis [25]. Photosynthesis was reduced by 25% per gram of fresh tissue weight in BYDV-infected plants [26]. Many studies have shown that virus infection can trigger severe chlorophyll breakdown within the host. An imbalance between biosynthesis and catabolic turnover of green pigments in plant tissues indicates profound inhibition of the photosynthesis process [27–30]. Phytohormone levels are also altered following BYDV infection. [31] undertook a detailed study of

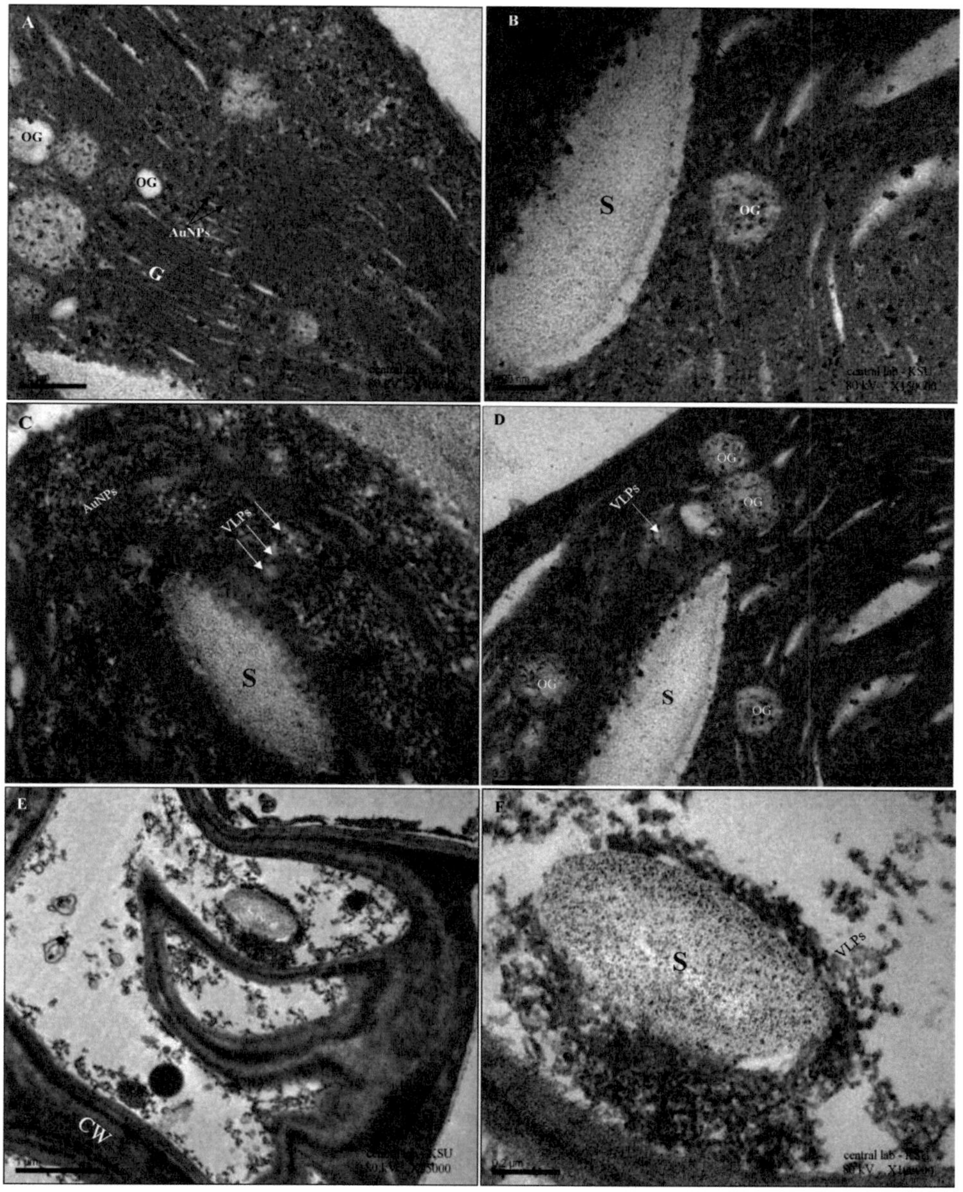

Figure 4.
Electron micrographs demonstrated many vesicles and starch body in the chloroplast. (**A**) *Noticed the grana was not striated and filled with light OG and having a lot of starch granules, Scale bar 0.2 μm.* (**B**) *Higher magnification from A, Scale bar 100 nm.* (**C**) *Starch granules surrounded by virus-like particles (VLPs) and gold nanoparticles (AuNPs) in the chloroplast in a highly magnified part of the grana, Scale bar 0.2 μm.* (**D**) *Micrograph of the grana decomposition in chloroplast showed the OG, which turned to hollow vesicles and filled with some cell material and VLPs from the virus-infected. (Outlined some aggregated VLPs), Scale bar 0.2 μm. Electron micrographs demonstrated the late stage of plasmolysis in the chloroplast.* (**E**) *Abnormal deformed invagination of the cell wall with remained starch body according to plasmolysis inside the cell. Scale bar 1 μm.* (**F**) *Plasmolysis of chloroplast with remained one starch granules surrounded by Nanoparticles associated with some VLPs., Scale bars 0.2 μm.*

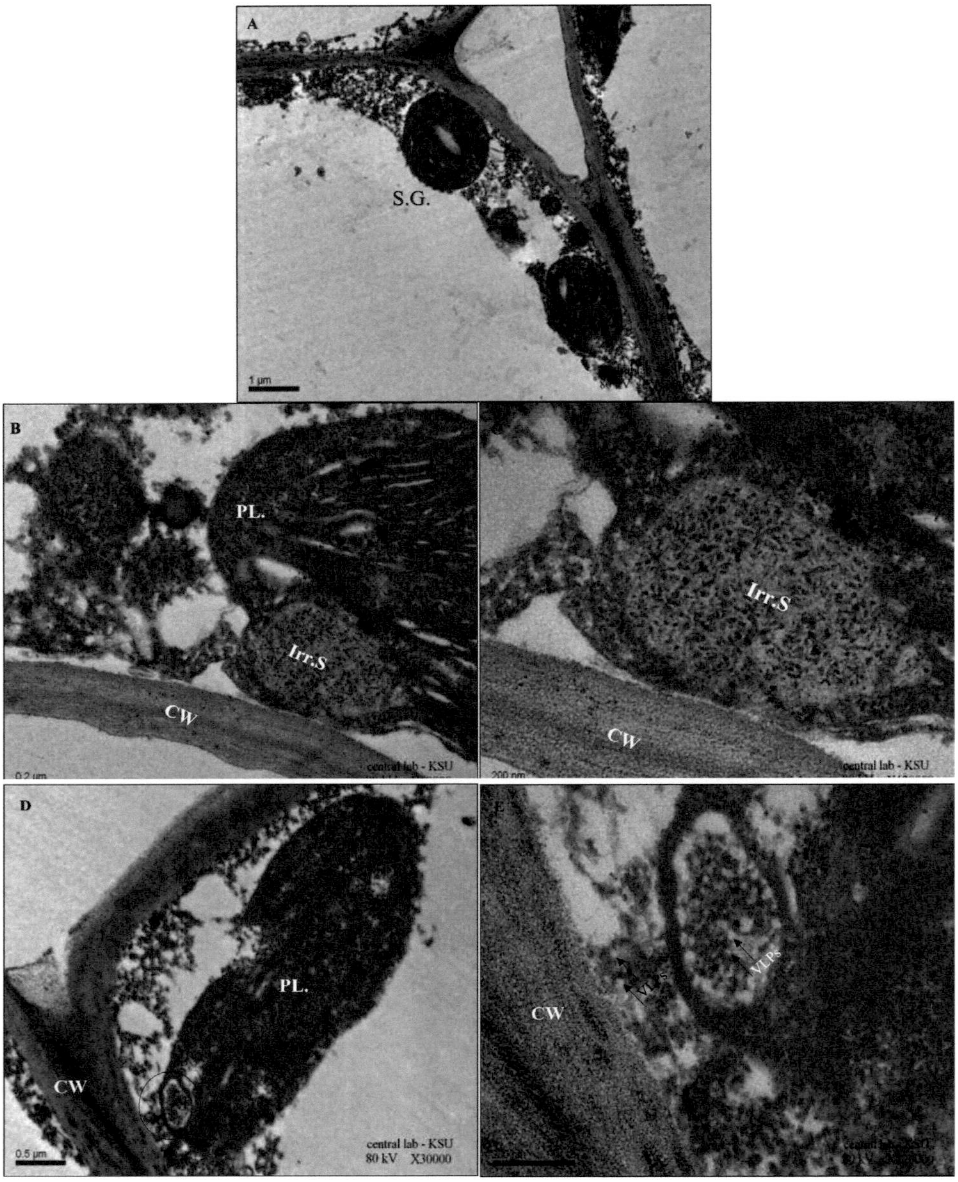

Figure 5.
Micrographs of different abnormalities in the plastid; rounded chloroplast. (**A**) *Micrograph of abnormally rounded chloroplast with one starch granules. Scale bar 1 µm. Micrographs of different abnormalities in the plastid; irregular starch granules* (**B**) *The cell contains an abnormal amorphous material adjacent both plastid and cell wall, Scale bars 0.2 µm* (**C**) *higher magnification showed this material having portentous contents with different irregular starch granules with many AuNPs. Scale bar 200 nm. Micrographs of different abnormalities in the plastid; deformed and decomposed chloroplast.* (**D**) *Elongated and tiered in the chloroplast with (OG) and have some hollow (circle outlined) surrounded part near the border of the plastid, some VLPs were located beside these hollow, Scale bar 0.5 µm.* (**E**) *a higher magnification illustrated the content of vacuole having many VLPs inside, outside and near the cell wall (arrow) indicted VLPs, Scale bar 200 nm.*

abscisic acid, jasmonic acid, methyl jasmonate, methyl salicylate (MS), and salicylic acid (SA) following BYDV infection. They followed the phytohormones over time and with different watering conditions than undamaged controls and seedlings infested with non-viruliferous aphids. Total hormone concentrations in BYDV-infected plants were more significant than those in sham-treated and control plants. SA was higher in infected plants, but MS was low.

Turnip yellow mosaic virus (TYMV) replication machinery interacts with the outer membrane of infected cell chloroplasts [32], like in our study in **Figure 5(A, B, D,** and **E)**. In TYMV-infected cells, the chloroplasts adopt a cup-shaped form aggregate and swell. Two TYMV proteins are targeted to the chloroplast membrane and are involved in these processes, the 66 and 140 kDa proteins. The 140 kDa protein induces chloroplast aggregation and induces the invagination of the outer chloroplast membrane to form small peripheral vesicles, which are the sites of TYMV replication. Invaginations of endoplasmic and the plastid membrane were precise in our micrographs; **Figures 2(A)** and **(B)** and **4(A)**.

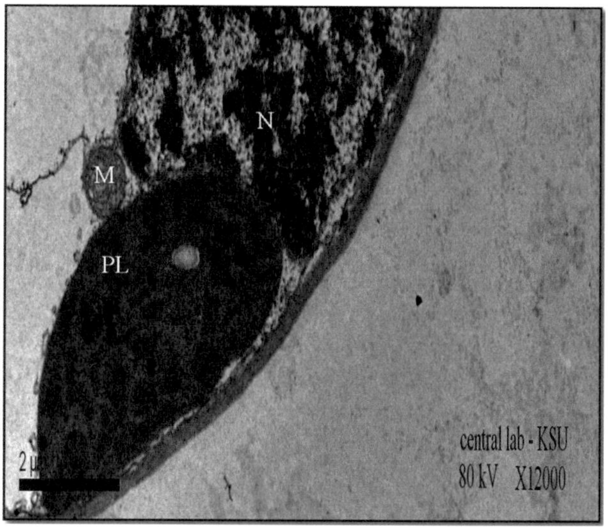

Figure 6.
The mitochondrion is a semi-autonomous double-membrane-bound organelle in this figure. Mitochondria are commonly measured between 0.75 and 3 μm² in the area but vary considerably in size and structure. The organelle is composed of compartments; these compartments include the outer membrane, the intermembrane space, the inner membrane, and the cristae and matrix. Scale bar 2 μm.

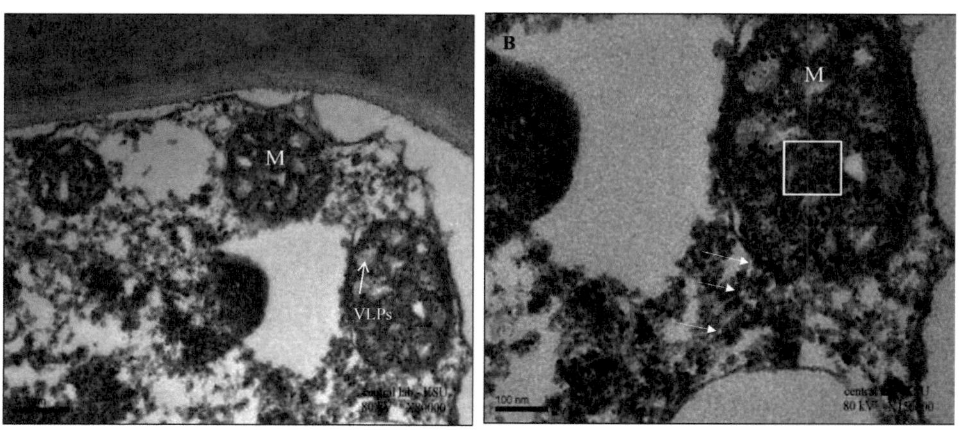

Figure 7.
*Micrographs of a distorted mitochondrion with internal membrane-bound areas containing amorphous inclusion bodies. (**A and B**) A lot of amorphous inclusion bodies (outlined) inside and outside the mitochondria with virus-like particles VLPs (arrows), Scale bars 0.2 μm, 100 nm.*

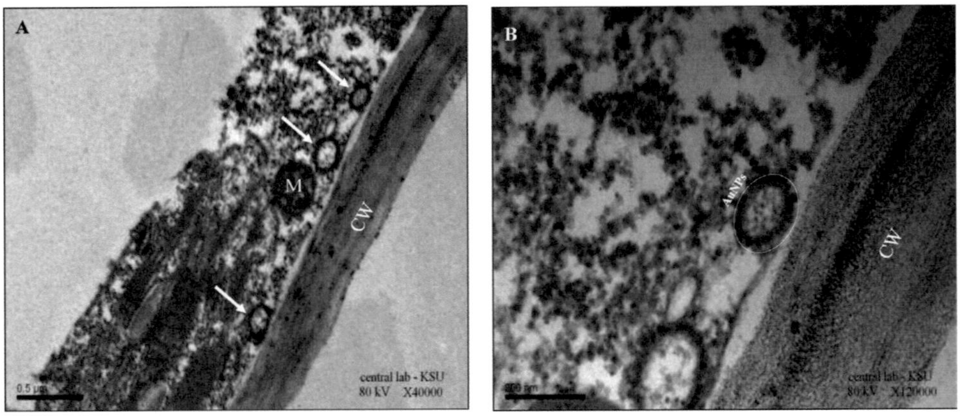

Figure 8.
Micrographs of localized and precipitated AuNPs around the mitochondria as shown in figure (**A**) *with fewer compartments than the health and higher magnification of the deformed mitochondrial organ and the AuNPs exist inside and along the cell wall as shown in figure* (**B**). *Scale bars 0.2 μm and 200 nm.*

Mitochondria and plastids in infected cells became degenerate in a way similar to those in uninfected maturing sieve elements. Negative staining preparation for Electron Microscopy is used for staining virus particles and the morphological and cytological side of healthy (**Figure 6**) and treated leaves [33]. Some alterations were detected in mitochondria of fully infected phloem cells, **Figure** 7 (**A**) and (**B**). Furthermore, virus-specific vesicles with the limiting membrane were developed in the cytoplasm.

Deformed mitochondrial organ with fewer compartments than the healthy and the AuNPs exist inside and along the cell wall, **Figure** 8(**A**) and (**B**).

Abbreviations

BYDV-PAV	Barley Yellow Dwarf Virus
BYDV-GAV	Barley Yellow Dwarf Virus
ROS	Reactive Oxygen Species
ABA	Abscisic acid
ET	Ethylene
TMV	Tobacco mosaic virus
CP	Coat protein
EM	Electron microscope
TEM	Transmission electron microscope
AuNPs	Gold nanoparticles
TIO_2	Titanium Oxide
RCA	Rubisco activase
mRNA	Messenger Ribonucleic Acid
TiO_2NPs	Nano-anatase particles
MS	Methyl salicylate
SA	Salicylic acid
TYMV	Turnip yellow mosaic virus
KDA	kilodalton
S	Starch
OG	Osmophilic globule
MLB	Myelin like bodies
G	Grana

DOI: http://dx.doi.org/10.5772/intechopen.97440

VLPs	Virus-like particles
CW	Cell wall
SG	Starch granules
IRR.S	Irregular starch granules
N	Nucleus
M	Mitochondria
PL	Plastid

Author details

Noorah Abdulaziz Othman Alkubaisi[1*] and Nagwa Mohammed Amin Aref[2]

1 Department of Botany and Microbiology, College of Science, King Saud University, Riyadh, Kingdom of Saudi Arabia

2 Department of Microbiology, College of Agriculture, Ain Shams University, Cairo, Egypt

*Address all correspondence to: nalkubaisi@ksu.edu.sa

References

[1] S. Choudhury, H. Hu, H. Meinke, S. Shabala, G. Westmore, P. Larkin, M. Zhou, Barley yellow dwarf viruses: infection mechanisms and breeding strategies, Euphytica 213(8) (2017) 168.

[2] C. Gill, J. Chong, Development of the infection in oat leaves inoculated with barley yellow dwarf virus, Virology 66(2) (1975) 440-453.

[3] T. Zhou, A.M. Murphy, M.G. Lewsey, J.H. Westwood, H.-M. Zhang, I. Gonzalez, T. Canto, J.P. Carr, Domains of the cucumber mosaic virus 2b silencing suppressor protein affecting inhibition of salicylic acid-induced resistance and priming of salicylic acid accumulation during infection, The Journal of general virology 95(Pt 6) (2014) 1408.

[4] C. D'Arcy, L. Domier, M. Mayo, Family luteoviridae, Virus Taxonomy: VIIIth Report of International Committee on Taxonomy of Viruses (2005).

[5] K.H. Orwin, M.U. Kirschbaum, M.G. St John, I.A. Dickie, Organic nutrient uptake by mycorrhizal fungi enhances ecosystem carbon storage: a model-based assessment, Ecology Letters 14(5) (2011) 493-502.

[6] Y. Wang, R. Chen, Y. Hao, H. Liu, S. Song, G. Sun, Transcriptome analysis reveals differentially expressed genes (DEGs) related to lettuce (Lactuca sativa) treated by TiO 2/ZnO nanoparticles, Plant Growth Regulation 83(1) (2017) 13-25.

[7] W. Yu, H. Xie, Y. Li, L. Chen, Experimental investigation on thermal conductivity and viscosity of aluminum nitride nanofluid, Particuology 9(2) (2011) 187-191.

[8] F. Gao, F. Hong, C. Liu, L. Zheng, M. Su, X. Wu, F. Yang, C. Wu, P. Yang, Mechanism of nano-anatase TiO 2 on promoting photosynthetic carbon reaction of spinach, Biological trace element research 111(1-3) (2006) 239-253.

[9] X. Ma, R. Jian, P.R. Chang, J. Yu, Fabrication and characterization of citric acid-modified starch nanoparticles/plasticized-starch composites, Biomacromolecules 9(11) (2008) 3314-3320.

[10] M. Qi, Y. Liu, T. Li, Nano-TiO 2 improve tomato leaves' photosynthesis under mild heat stress, Biological trace element research 156(1-3) (2013) 323-328.

[11] M. Nasrollahzadeh, M.S. Sajadi, M. Atarod, M. Sajjadi, Z. Isaabadi, An Introduction to Green Nanotechnology, Academic Press2019.

[12] A.O. Govorov, I. Carmeli, Hybrid structures composed of the photosynthetic system and metal nanoparticles: plasmon enhancement effect, Nano letters 7(3) (2007) 620-625.

[13] E. Cabiscol, J. Ros, Oxidative damage to proteins: structural modifications and consequences in cell function, Redox proteomics: from protein modification to cellular dysfunction and disease (2006) 399-471.

[14] V. Chinnusamy, J.-K. Zhu, Epigenetic regulation of stress responses in plants, Current opinion in plant biology 12(2) (2009) 133-139.

[15] S.T. Thul, B.K. Sarangi, Implications of nanotechnology on plant productivity and its rhizospheric environment, Nanotechnology and plant Sciences, Springer2015, pp. 37-53.

[16] M.H. Siddiqui, M.H. Al-Whaibi, M. Firoz, M.Y. Al-Khaishany, Role of nanoparticles in plants, Nanotechnology and Plant Sciences, Springer 2015, pp. 19-35.

[17] N. Verbruggen, C. Hermans, H. Schat, Molecular mechanisms of metal hyperaccumulation in plants, New phytologist 181(4) (2009) 759-776.

[18] J.A. Lemire, J.J. Harrison, R.J. Turner, Antimicrobial activity of metals: mechanisms, molecular targets, and applications, Nature Reviews Microbiology 11(6) (2013) 371-384.

[19] V. Kumar, P. Guleria, V. Kumar, S.K. Yadav, Gold nanoparticle exposure induces growth and yield enhancement in Arabidopsis thaliana, Science of the total environment 461 (2013) 462-468.

[20] S. Arora, P. Sharma, S. Kumar, R. Nayan, P. Khanna, M. Zaidi, Gold-nanoparticle induced enhancement in growth and seed yield of Brassica juncea, Plant Growth Regulation 66(3) (2012) 303-310.

[21] W. Rong, X. Wang, X. Wang, S. Massart, Z. Zhang, Molecular and ultrastructural mechanisms are underlying yellow dwarf symptom formation in wheat after infection of Barley Yellow Dwarf Virus, International Journal of molecular sciences 19(4) (2018) 1187.

[22] W.A. Miller, S. Dinesh-Kumar, C.P. Paul, Luteovirus gene expression, Critical Reviews in plant sciences 14(3) (1995) 179-211.

[23] S.G. Jensen, Metabolism and Carbohydrate Composition in Barley, Phytopathology 62 (1972) 587-592.

[24] W.E. Riedell, R.W. Kieckhefer, M.A. Langham, L.S. Hesler, Root and shoot responses to bird cherry-oat aphids and barley yellow dwarf virus in spring wheat, Crop Science 43(4) (2003) 1380-1386.

[25] S.G. Jensen, J.W. Van Sambeek, Differential effects of barley yellow dwarf virus on the physiology of tissues of hard red spring wheat, Phytopathology 62 (1972) 290-293.

[26] S. Jensen, P. Fitzgerald, J. Thysell, Physiology and Field Performance of Wheat Infected with Barley Yellow Dwarf Virus 1, Crop Science 11(6) (1971) 775-780.

[27] A. Marquardt, G. Scalia, P. Joyce, J. Basnayake, F.C. Botha, Changes in photosynthesis and carbohydrate metabolism in sugarcane during the development of Yellow Canopy Syndrome, Functional plant biology 43(6) (2016) 523-533.

[28] P. Klein, C.M. Smith, Host plant selection and virus transmission by Rhopalosiphum maidis are conditioned by potyvirus infection in Sorghum bicolor, Arthropod-Plant Interactions (2020) 1-13.

[29] I. Łukasik, A. Wołoch, H. Sytykiewicz, I. Sprawka, S. Goławska, Changes in the content of thiol compounds and the activity of glutathione s-transferase in maize seedlings in response to a rose-grass aphid infestation, PloS one 14(8) (2019) e0221160.

[30] H. Sytykiewicz, P. Czerniewicz, I. Sprawka, R. Krzyzanowski, Chlorophyll content of aphid-infested seedling leaves of fifteen maize genotypes, Acta Biologica Cracoviensia. Series Botanica 55(2) (2013).

[31] D.W. Mornhinweg, S. Ullrich, Biotic stress in barley: Insect problems and solutions, Barley: Production, improvement, and uses (2011) 355-390.

[32] D. Prod'homme, A. Jakubiec, V. Tournier, G. Drugeon, I. Jupin, Targeting the turnip yellow mosaic virus 66K replication protein to the chloroplast envelope is mediated by the 140K protein, Journal of virology 77(17) (2003) 9124-9135.

[33] J. Kuo, Electron microscopy: methods and protocols, Springer Science & Business Media 2007.

Chapter 4

Cell Wall

Noorah Abdulaziz Othman Alkubaisi and Nagwa Mohammed Amin Aref

Abstract

The application of AuNPs on the infected barley cultivar had great damage results on Barley Yellow Dwarf Virus (BYDV-PAV) particles in TEM. Observation of TEM images provided an insight into the transport of AuNPs through the plasmodesmata endoplasmic reticulum route, where they likely accumulated as the channels narrowed. The cytoplasmic parenchyma cell components do not have an intact peripheral location, but taking irregular shapes, internal movement between adjacent two cells seems to be the VLPs moved toward via plasmodesmata. TEM micrographs; showing different abnormalities in the cell wall due to viral infection. Application of AuNPs revealed sticky Integrated AuNPs inside the cell wall with low and high density. The mechanical transportation of the virus through the sieve elements with endosomes was observed. The mechanical transportation of virus particles through the cell wall with some vesicles, amorphous inclusions, and filamentous particles was proved through the sieve elements with filamentous strands.

Keywords: cell wall, Barley Yellow Dwarf Virus (BYDV-PAV), Gold nanoparticles (AuNPs), parenchyma cells, plasmodesmata, low density of AuNPs, AuNPs with high density, auNPs with high density, twisted cell wall, proteinous contents, endosomes, amorphous inclusions, mechanical transportation, filamentous particles, sieve elements

1. Introduction

The phloem-limited virus not only causes damage to the phloem and associated plasmodesmata [1] but also produces viral proteins that influence plasmodesmata structure to allow the passage of viral genomes or virus particles **Figure 1**. Compared to **Figure 2** of healthy cells. For example, the TMV MP accumulates in plasmodesmata and alters the size exclusion limit of the channel [2]. The application of AuNPs on the barley cultivar had great damage results on virus in TEM. The interaction performance in our plates in **Figures 3** and **4** had different degrees between VLPs and AuNPs, illustrated as clotting, surrounding, integration, and accumulation particles. AuNPs with sticky and low density in **Figure 3(A)–(D)** and with high dark density inside the twisted cell wall in **Figure 4(A)–(F)**. The DNA AuNPs can bind to the strand upon entry to the cell's cytoplasm, preventing translation of the mRNAs corresponding protein. As found in **Figure 5(A)** and **(B)**, which pronounced many filamentous strands bind with AuNPs crossing the cell wall, that could have the same influence [3]. Outside of medical technology, the deployment of gold nanoparticles in photovoltaics by embedding them into thin solar cells has increase solar energy conversion efficiencies. The mechanism of AuNPs inside the plant was illustrated [4]. It was recorded that the formation of Nanoparticles still needs more clarification,

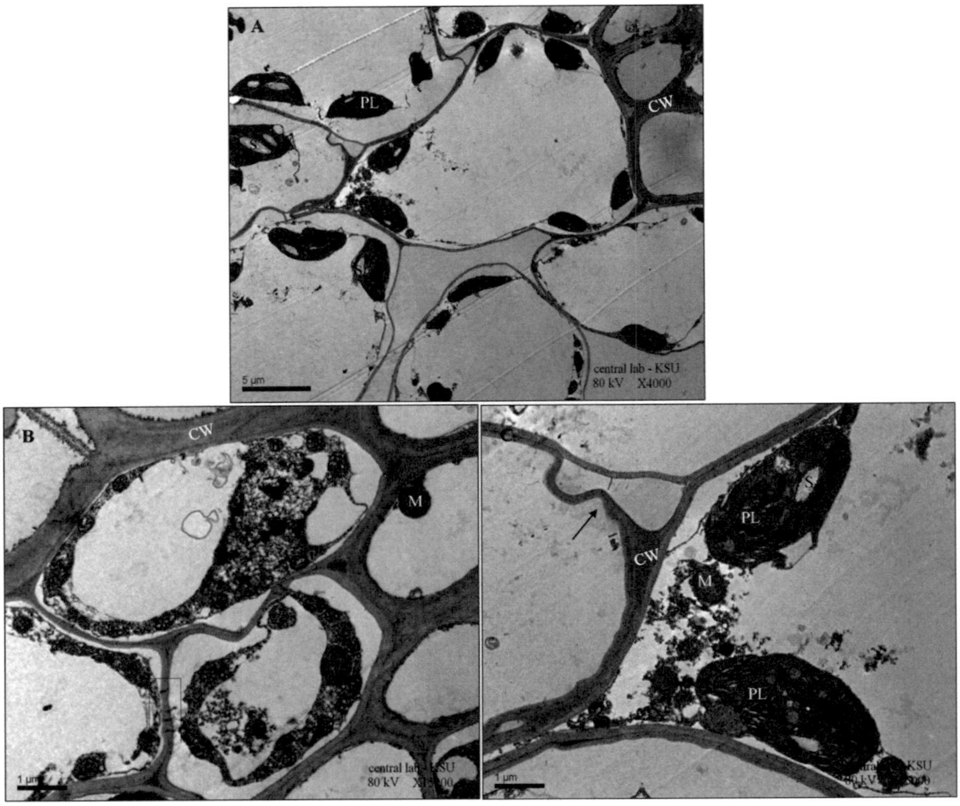

Figure 1.
Micrographs of different abnormalities in the cell wall due to viral infection. (A) Parenchyma cells of the entire contents, scale bar 5 µm. (B) The cytoplasmic components do not have an intact peripheral location, but taking irregular shapes internal movement between adjacent two cells, it seems to be the VLPs moved toward via plasmodesmata (outlined), scale bar 1 µm. (C) Increased of the cell wall thickness, irregular wavy cell wall (arrow). Scale bar 1 µm.

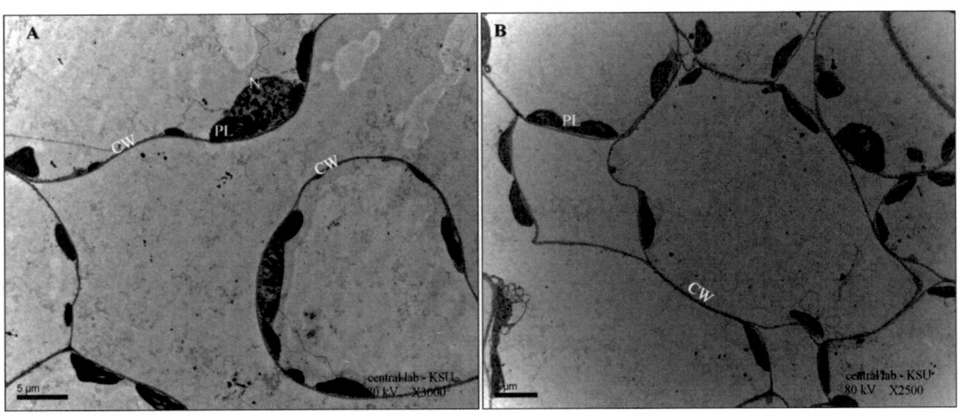

Figure 2.
Electron micrographs showing a general view. Healthy cells with different adjacent conducted with the cell wall, as shown in (A and B), scale bars 5 µm.

whether they are created outside and then translocated to plants or whether they are formed by the reduction of metal salts within the plants themselves. The uptake of Nanoparticles and translocation across root cells, in which several active and passive transport processes involved, depends on the type of metal ions and plant

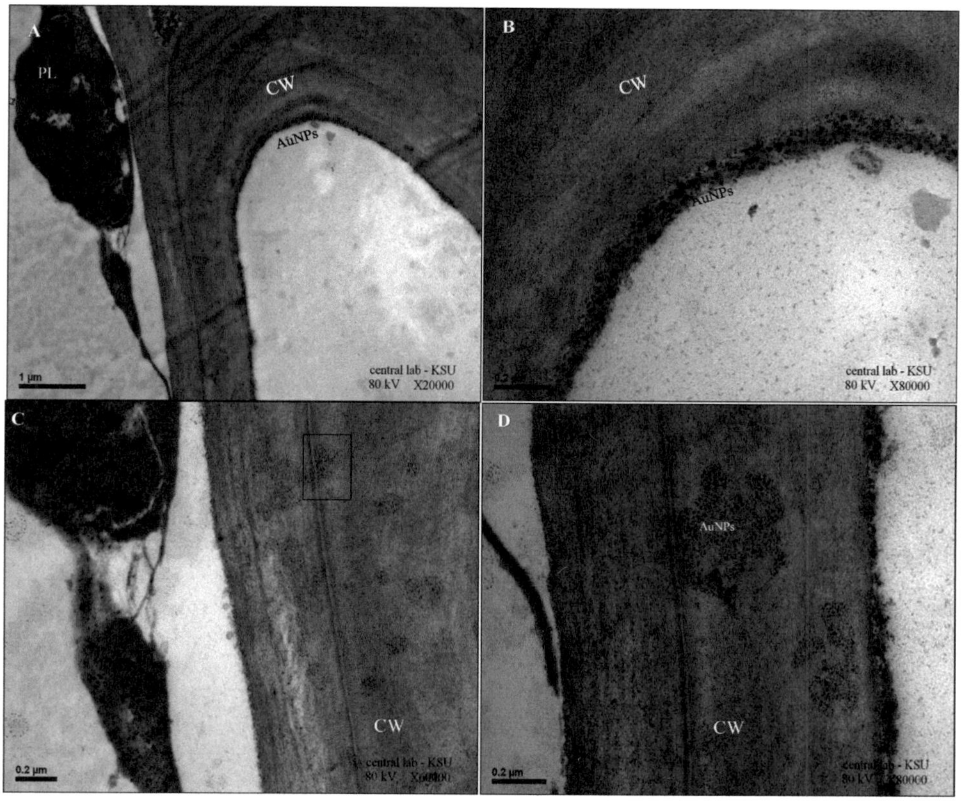

Figure 3.
Micrographs of the AuNPs sticky and integrated inside the cell wall with low density. (A) AuNPs were gathering together, having large irregular distribution mass net inside the cytoplasm, and circulated the inner side of the cell wall, scale bar 1 µm. (B) Enlarged magnification of A declares AuNPs in the inner border of the cell wall and cytoplasmic area. Scale bar 0.2 µm. (C) AuNPs spotted and attracted firmly on the cell wall (higher magnification of A), scale bar 0.2 µm. (D) Distribution of the AuNPs (higher magnification of C), scale bar 0.2 µm.

species, **Figure 5(A)**, **(B)**, **(E)**, and **(L)**. The amount of Nanoparticle accumulation in plants also varies with the reduction potential of ions and the lowering capacity of plants that depends on the presence of various polyphenols and other heterocyclic compounds present in plants, as shown in **Figure 3(C)** and **(D)** [5] that sieving properties are determined by the pore diameter of cell wall ranging from 5 to 20 nm **Figure 5(L)**.

2. Nanoparticles, interalization and transportation inside the cell wall

During endocytosis, further internalization occurs with a cavity-like structure that forms around the nanoparticles by the plasma membrane. They may, besides, cross the membrane using embedded transport carrier proteins or through ion channels. A band of AuNPs surrounded the cell adjacent and inside the cell wall, as **Figure 5(C)**.

Nuclear Localization Signal Peptide, NLS peptide, is a virus-derived protein fragment [6] that interacts with intracellular proteins for transport across the nuclear envelope. Interpreted that NLS peptides to the DNA AuNP did alter the intercellular localization of the AuNPs [7] but did not change the binding properties of the oligonucleotides (**Figure 5(I)**, **(K)** and **(M)**.

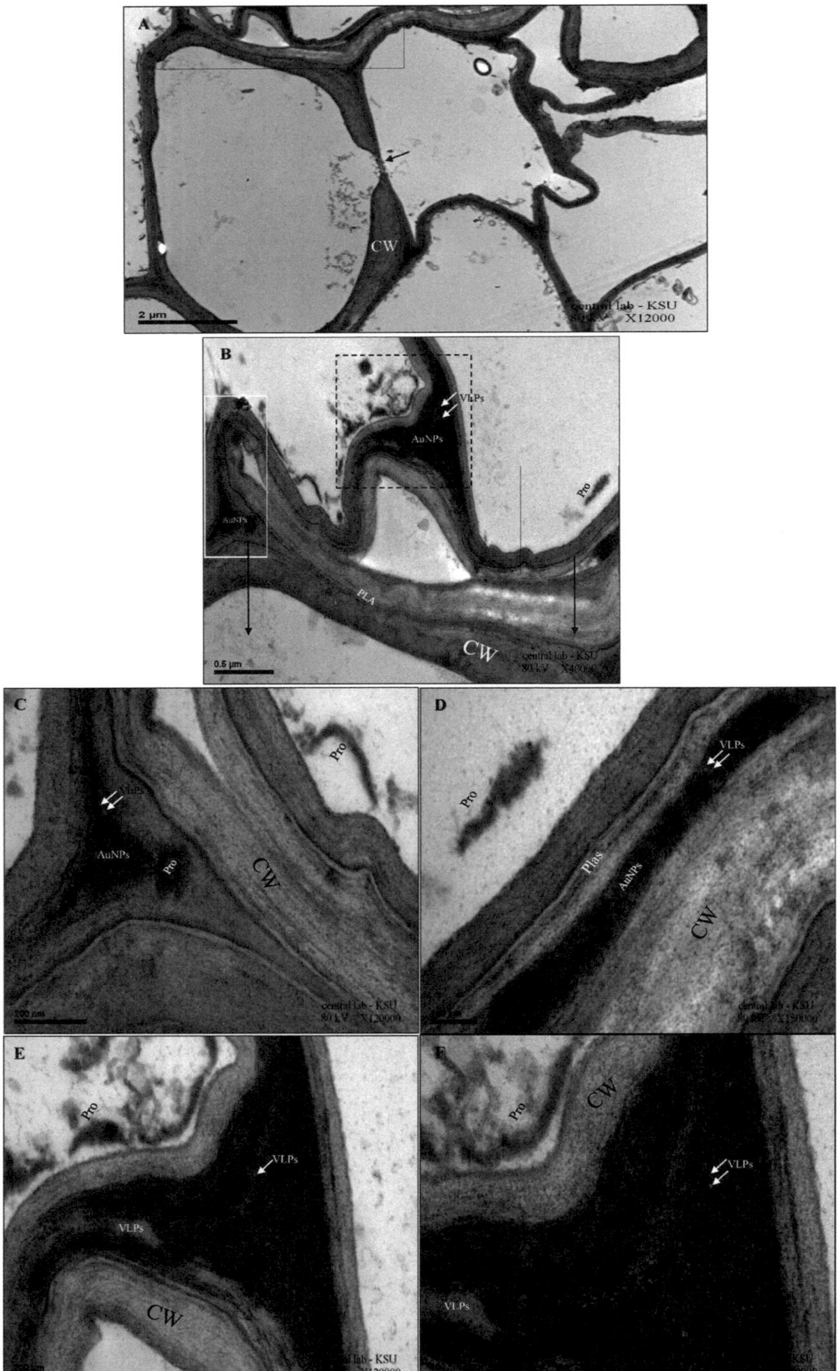

Figure 4.

Micrographs of the AuNPs integrated inside the cell wall with high density. (A) The cell wall had twisted invaginated shapes as shown in the bold box, and some dynamic move the proteinous material via plasmodesmata (arrow), scale bar 2 µm. (B) Very twisted cell wall as of accumulated dark density AuNPs inside the interaction of their cell wall structure joined with VLPs in three different parts, scale bar 0.5 µm. (C) The AuNPs had low concentration and density with VLPs (white box in B) than in (dashed box in B), scale bar 200 nm. (D) Higher magnification showed high accumulation density of AuNPs with VLPs had very dark color along extended the cell wall in the middle of it (dashed box in B), proteinous contents near the cell wall on the right side (black box in B), scale bar 100 nm. (E and F) VLPs trapped in massive of AuNP inside the cell wall, with higher resolutions, scale bars 200 nm,100 nm.

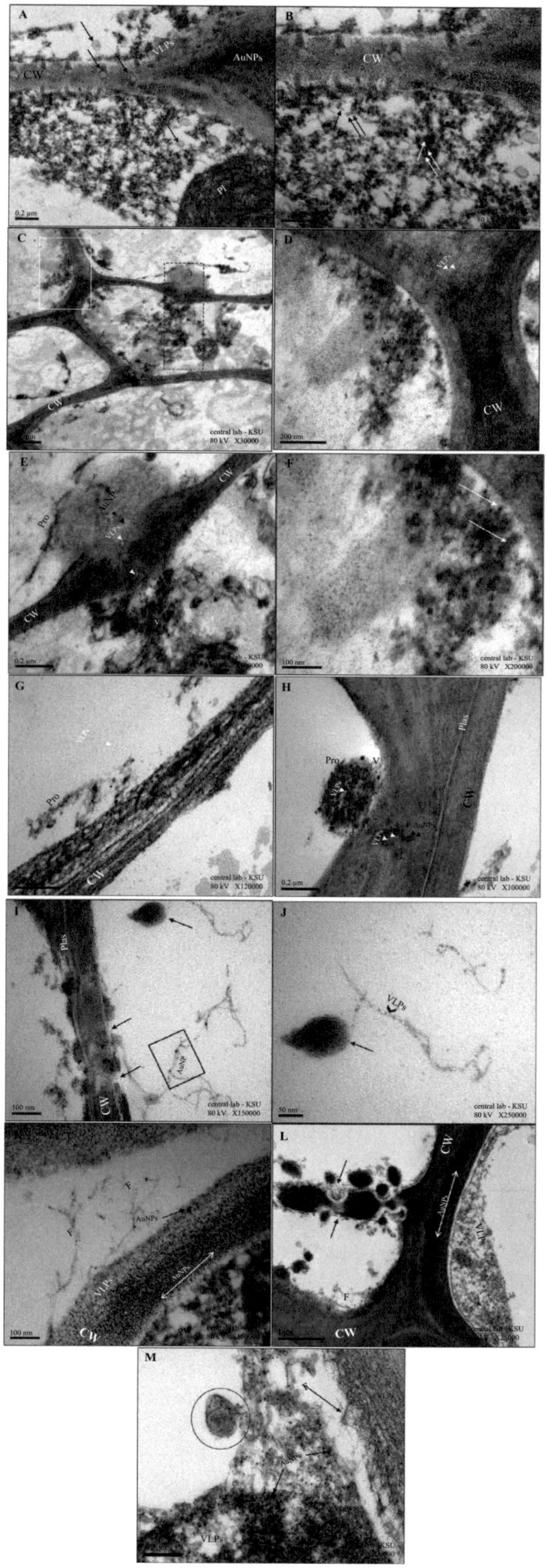
A
VLPs
AuNPs
CW
Pl
0.2 µm
B
CW
C
CW
central lab - KSU
80 kV X30000
D
VLPs
AuNPs
CW
200 nm
E
CW
Pro
AuNPs
VLPs
CW
0.2 µm
F
100 nm
central lab - KSU
80 kV X200000
G
VLPs
Pro
CW
central lab - KSU
80 kV X120000
H
Plas
Pro
V
VLPs
CW
AuNPs
0.2 µm
central lab - KSU
80 kV X100000
I
Plas
AuNPs
CW
100 nm
central lab - KSU
80 kV X150000
J
VLPs
50 nm
central lab - KSU
80 kV X250000
F
AuNPs
F
VLPs
AuNPs
CW
100 nm
L
CW
AuNPs
VLPs
F
CW
M
VLPs

Figure 5.
Micrographs of the mechanical transportation of virus through the sieve elements with endosomes. (A) Viral particles associated with AuNPs with endosomes in plasmodesmata connecting with two adjacent parenchyma cells, mechanical transport movement of some vesicles and amorphous inclusions, and filamentous particles stuck or near the cell wall, scale bar 0.2 μm. (B) Mechanical transportation of some blowing VLPs in and out through the cell wall and cytoplasmic membrane compartment (arrows), Nemours AuNPs attracted with deformed VLPs in the cytoplasm, cell wall, and PL. Some empty endosomes exist inside the cytoplasm in the inner part of the cells (arrow), scale bar 0.2 μm. Micrographs of the mechanical transportation of virus through the cell wall with some vesicles, amorphous inclusions, and filamentous particles. (C) The cell is full of many different kinds of abnormal structure endosomes (dashed box) as well as many accumulated and irregular distribution of light and dark area (white box), scale bar 0.5 μm. (D and F) showed the aggregation of VLPs with dark AuNPs near to the cell wall that has intermediate virus particles in the dark area (arrows). Two neighboring cells separated with the cell wall, some cytoplasmic material crossing the cell wall by one of that endosomes to the other cell with VLPs (arrows). Scale bars 200,100 nm. (E) Mechanical transport movement of some vesicles and amorphous inclusions and filamentous particles stuck or near the cell wall. (higher magnification of D), scale bar 0.2 μm. Micrographs of the mechanical transportation of virus through the sieve elements with filamentous strands. (I) Crystalline array (arrow) and some filamentous strands (outlined) in the cytoplasm joined with dark AuNPs; the cell wall has two locations of pits that seem to be the gate of moving materials (arrows), scale bar 100 nm. (J) High magnification for the pseudo-crystal array (arrow) associated with light VLPs (arrow), scale bar 50 nm. (K) Numerous filamentous inclusions extended in the cell lumen toward the cell wall interspersed with AuNPs and VLPs, scale bar100nm. (L) Parenchyma cell starts to divide, de-generated cell wall and abnormal deformed cell wall inside the cell (arrows), AuNPs extended along the complete cell wall in high density (arrow with two directions), scale bar 1 μm. (M) Illustrate the endosomes (circular) and filamentous strands near the cell arrow indicates, VLPs distributed in the right side and middle of the figure while the AuNPs concentrated in the lower part inside the cytoplasm. Scale bar 0.2 μm.

Two-photon excitation microscopy was used to monitor MWCNTs piercing the cell wall of wheat roots and reaching the cytoplasm [8] without totally entering the cell. After root uptake and penetration of the epidermal cells of Engineered nanomaterials ENMs, transport demand circulation across the root and the xylem. ENMs transported through cell wall pores, the apoplastic pathway, or the symplastic pathway through plasmodesmata, channels that connect neighboring cells around 40 nm in diameter [9]. The uptake and translocation of ENMs in plants are not only related to the particle composition, size, shape, surface properties, but also the type of plant species. Positively charged AuNPs were most readily taken up by plant roots [10], while negatively charged AuNPs most efficiently translocated into stems and leaves from the roots. Higher amounts of the AuNPs accumulated in Radish and ryegrass roots generally than rice and pumpkin roots, as we found in wheat in our study. Utilized AuNPs in that study aggregated to statistically considerable extents in rice shoots; however, none of them depleted in the shoots of radishes and pumpkins. Near, the tissue scale uptake and locative distribution NPs in rice roots and shoots were influenced by the surface charges of Au NPs [11]. Au concentration in rice roots followed the order of AuNP (+) > AuNP (0) > AuNP (−) but with the reversed order for shoots, indicating preferential translocation of negatively charged Au NPs. AuNPs in the xylem within the leaves indicated that the NPS was transported during nutrients and water uptake [12].

Direct penetration of NPS through the cell wall can be envisioned for smaller-size NPS as the pore size on the cell wall (2–20 nm) may limit the passage to the NPs more significant than 20 nm [12]. Furthermore, the cell membrane acts as yet another barrier for an extraneous agent to pass through. To test this hypothesis, [13] investigated uptake and distribution of 3.5 nm or 18 nm-sized citrate-coated AuNPs in tobacco (Nicotiana Xanthi), where the authors observed uptake of Au only from 3.5 nm AuNPs treatments [13], distinct with 18 nm-sized AuNPs treatments, and the larger-sized particles adhered to tobacco root surfaces. Also, exposure to 3.5 nm-sized AuNPs resulted in leaf necrosis sustain in plant death. It did not occur with 18 nm AuNPs treatments [13]. On the other hand, a size threshold may happen for NPS translocation to the leaves, which they mentioned to be <36 nm; meanwhile, the accumulation of TiO_2NPs in the wheat root could only exist if NPs are <140 nm in diameter, with higher aggregation that take place when NPs were much smaller (in the size range 14–22 nm) [14]. Combined, these observations from chemically different ENMs bolster the premise that particle size could be an essential factor regulating ENM bio uptake in plants.

Abbreviations

VLPs	Virus-like particles
TMV	Tobacco mosaic virus
MP	Movement protein
AuNPs	Gold nanoparticles
TEM	Transmission electron microscope
DNA	Deoxyribonucleic Acid
mRNAs	Messenger Ribonucleic Acid
MWCNTs	Multi-walled carbon nanotubes
NLS	Nuclear Localization Signal Peptide;
ENMs	Engineered nanomaterials
NPS	Nanoparticles
nm	NanoMeter
PL	Plastid
CW	Cell Wall
M	Mitochondria
S	Starch
(BYDV-PAV)	Barley Yellow Dwarf Virus
PLA	Plasmodesmata
PRO	Proteinous
F	Filamentous
V	Vesicles

Author details

Noorah Abdulaziz Othman Alkubaisi[1]* and Nagwa Mohammed Amin Aref[2]

1 Department of Botany and Microbiology, College of Science, Female Scientific and Medical Colleges, King Saud University, Riyadh, Kingdom of Saudi Arabia

2 Faculty of Agriculture, Department of Microbiology, Ain Shams University, Muhafazat al Qahirah, Cairo, Egypt

*Address all correspondence to: nalkubaisi@ksu.edu.sa

References

[1] A. Comeau, S. Haber, Breeding for BYDV tolerance in wheat as a basis for a multiple stress tolerance strategy, Barley Yellow Dwarf Disease: Recent Advances and Future Strategies (2002) 82.

[2] E. Neumann, A. Sprafke, E. Boldt, H. Wolf, Biophysical considerations of membrane electroporation, Guide to electroporation and electrofusion (1992) 77-90.

[3] K.N. Thakkar, S.S. Mhatre, R.Y. Parikh, Biological synthesis of metallic nanoparticles, Nanomedicine: nanotechnology, biology, and medicine 6(2) (2010) 257-262.

[4] J. Nari, G. Noat, JJBJ Ricard, Pectin methylesterase, metal ions, and plant cell-wall extension. Hydrolysis of pectin by plant cell-wall pectin methylesterase, 279(2) (1991) 343-350.

[5] A. Fleischer, M.A. O'Neill, R. Ehwald, The pore size of non-graminaceous plant cell walls is rapidly decreased by borate ester cross-linking of the pectic polysaccharide rhamnogalacturonan II, Plant Physiology 121(3) (1999) 829-838.

[6] H.-M. Chen, L.-T. Chen, K. Patel, Y.-H. Li, D.C. Baulcombe, S.-H. Wu, 22-Nucleotide RNAs trigger secondary siRNA biogenesis in plants, Proceedings of the National Academy of Sciences 107(34) (2010) 15269-15274.

[7] DS. Goldfarb, J. Gariépy, G. Schoolnik, R.D. Kornberg, Synthetic peptides as nuclear localization signals, Nature 322(6080) (1986) 641-644.

[8] E. Wild, K.C. Jones, Novel method for the direct visualization of in vivo nanomaterials and chemical interactions in plants, Environmental Science & technology 43(14) (2009) 5290-5294.

[9] LG Tilney, T.J. Cooke, P.S. Connelly, MS Tilney, The structure of plasmodesmata as revealed by plasmolysis, detergent extraction, and protease digestion, Journal of Cell Biology 112(4) (1991) 739-747.

[10] Z.-J. Zhu, H. Wang, B. Yan, H. Zheng, Y. Jiang, O.R. Miranda, V.M. Rotello, B. Xing, R.W. Vachet, Effect of surface charge on the uptake and distribution of gold nanoparticles in four plant species, Environmental Science & Technology 46(22) (2012) 12391-12398.

[11] J. Koelmel, T. Leland, H. Wang, D. Amarasiriwardena, B. Xing, Investigation of gold nanoparticles uptake and their tissue level distribution in rice plants by laser ablation-inductively coupled-mass spectrometry, Environmental pollution 174 (2013) 222-228.

[12] G. Zhai, KS. Walters, D.W. Peate, P.J. Alvarez, J.L. Schnoor, Transport of gold nanoparticles through plasmodesmata and precipitation of gold ions in woody poplar, Environmental science & technology letters 1(2) (2014) 146-151.

[13] T. Sabo-Attwood, J.M. Unrine, J.W. Stone, C.J. Murphy, S. Ghoshroy, D. Blom, P.M. Bertsch, L.A. Newman, Uptake, distribution and toxicity of gold nanoparticles in tobacco (Nicotiana Xanthi) seedlings, Nanotoxicology 6(4) (2012) 353-360.

[14] C. Larue, J. Laurette, N. Herlin-Boime, H. Khodja, B. Fayard, A.-M. Flank, F. Brisset, M. Carriere, Accumulation, translocation and impact of TiO2 nanoparticles in wheat (Triticum aestivum spp.): influence of diameter and crystal phase, Science of the total environment 431 (2012) 197-208.

Chapter 5

Cytoplasmic Matrix and Viroplasms Inclusions in the Presence of Gold Nanoparticles (AuNPs)

Noorah Abdulaziz Othman Alkubaisi and Nagwa Mohammed Amin Aref

Abstract

Cellular ultrastructure micrographs revealed striking changes resulting from the Barley Yellow Dwarf Virus (BYDV-PAV) infection in Electron microscopy. In the cytoplasm, the Gold nanoparticles (AuNPs) may bind with different cytoplasmic organelles and interfere with the treated site's metabolic processes. The micrographs of the treated plant leave with AuNPs showing; Endosomes, amorphous bodies, slender filaments fibers, myelin bodies with a high concentration of virus particles, and Gold Nanoparticles distributed in a circulated shape in the cytoplasm with virus particles.

Keywords: Cytoplasmic matrix, Barley Yellow Dwarf Virus (BYDV-PAV), Gold nanoparticles (AuNPs), Endosomes, Amorphous bodies, Slender filaments fibers, Myelin bodies

1. Introduction

In subsequent stages, significant accumulations of virus particles as in **Figures 1(E)** and **2(D)** and filaments were visible outside the nucleus throughout the cytoplasm, as in **Figure 1(D)**. When the virus was common throughout the cytoplasm, vesicles with fibrils were often still clustered around vacuole like areas in the cytoplasm, **Figure 1(A)–(C)**.

In addition to that, amorphous cellular inclusion bodies, was shown in **Figures 1(B)** and **2(A)–(C)**, numerous filamentous shapes appeared in **Figure 1(E)**. Endosomes as in **Figure 2(A)–(D)**, deformed invagination of the cell wall, **Figure 2(A)** virus localization, and cytopathic alterations of barley *Hordeum vulgare L.* were studied in root tissues infected with BYDV in root phloem tissues examined seven days after inoculation [1]. The virus was restricted to the phloem tissues, i.e., sieve elements, companion, and phloem parenchyma cells. Virus progeny was observed in the nucleus at an early stage of infection. No or few virions were observed in the cytoplasm at this stage, and cell organelles remained normal in their appearance. Uptake and presence of AuNPs in the cytoplasm various organelles of root and leaf cells of poplar plant by transmission electron microscopy were observed [2] and measured by inductively coupled plasma mass spectrometry (ICP-MS), **Figure 3(A)** and **(B)**.

IntechOpen

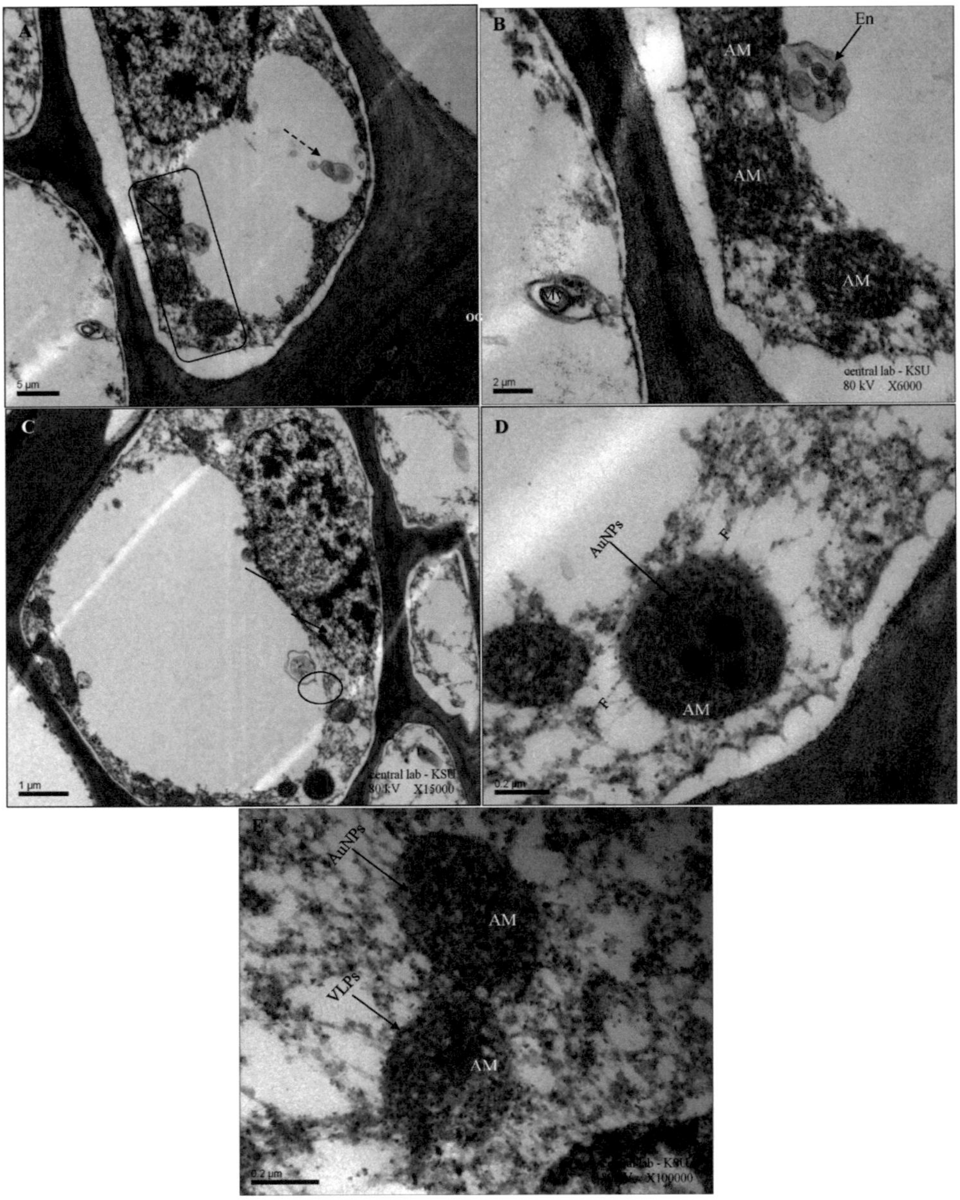

Figure 1.
Endosomes, amorphous bodies, and slender filaments fibers. *(A and C) Micrographs of general view for one cell contain an amorphous body (box outlined), myelin bodies (dashed arrow), and endosomes (arrow). Scale bars 5 µm, 1 µm. (B) Higher magnification of A indicated one endosome crossing the cell wall from the neighbouring cells to the other cell with VLPs. Scale bars 2 µm. (D) Higher magnification of C shows some slender filament fibers and VLPs with AuNPs near to amorphous bodies, Scale bar 0.2 µm. (E) smooth-surfaced structure with numerous AuNPs, Scale bar 0.2 µm.*

2. Viroplasms and gold nanoparticles

Inclusions of virus-derived material in virus-infected cells termed viroplasms. It may be quasicrystalline, amorphous, or crystalline, in appearance the accumulation of nascent virions in the host cell from some inclusions aggregates of viral proteins, **Figures 1(D)** and **(E)** and **2(A)–(C)**. Large aggregates of convoluted,

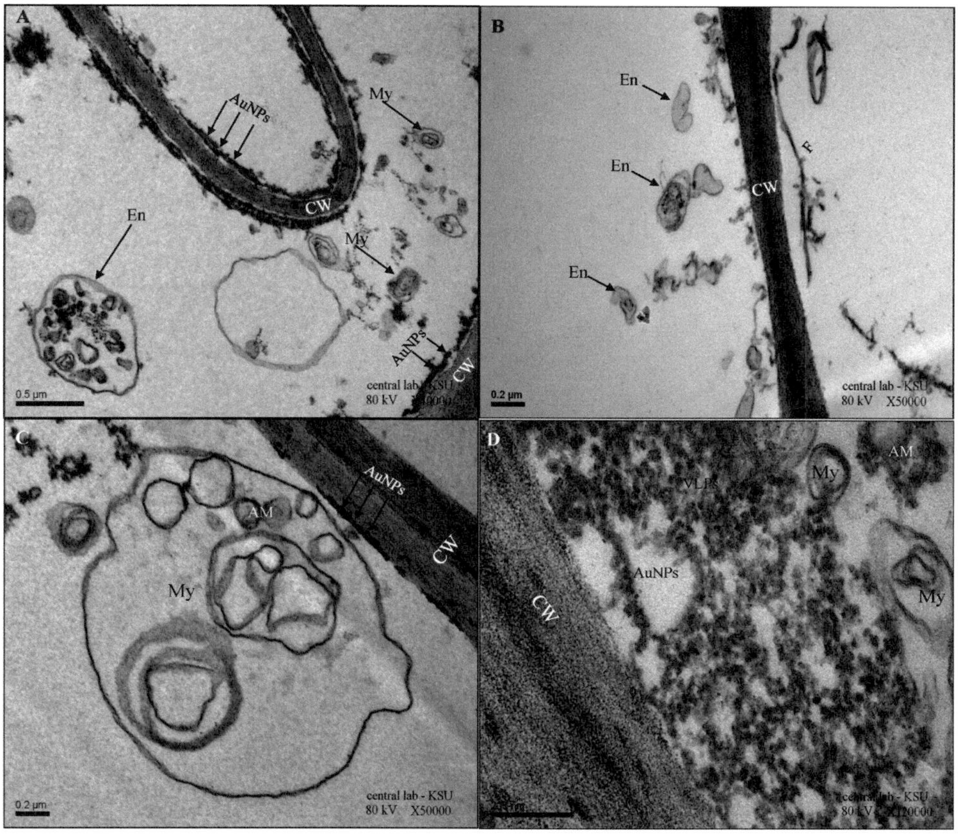

Figure 2.
Endosomes and myelin bodies with a high concentration of virus particles. *(A) Numerous endosomes (arrow) have AuNPs distributed in the cytoplasm of different sizes. Compact multi membrane structure (arrow) decorated in and out with a dense layer of AuNPs. Scale bar 0.5 µm. (B) The area around the cell wall in the cell endosomes along the cell wall with different sizes (arrows), filamentous shape particles in the other sides of the cell wall. Scale bar 0.2 µm. (C) Amorphous inclusion materials are surrounded by the bundle sheath myelin body (My). Scale bar 0.2 µm. (D) Micrograph of cellular myelin body trapped with both AuNPs and VLPs near the cell wall. Scale bar 200 nm.*

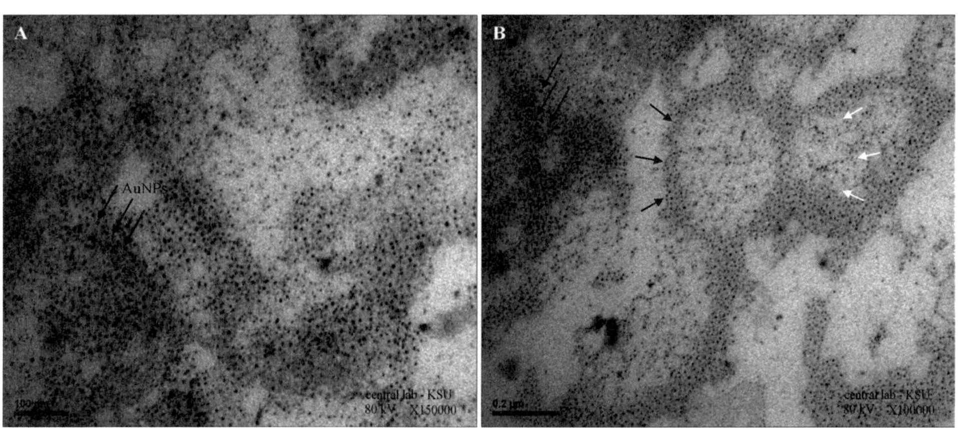

Figure 3.
Gold Nanoparticles distributed in a circulated shape in the cytoplasm with virus particles. *(A and B) Showed the VLPs (white arrows) and AuNPs (black arrows) gathering in the cytoplasm in circulated unregulated shapes having borders occupying the whole lumen of the cytoplasm. Scale bar 100 nm and 0.2 µm.*

branched, tubular bodies and myelin bodies as in **Figures 1(B)** and **2(A)**, **(C)**, and **(D)**, assumed to originate from the ER., were also observed in some parts of the cytoplasm. A densely staining amorphous material occurred both within and outside the tubules. Vesicles containing densely staining amorphous material or without any apparent contents were occasionally seen within these aggregates but were more common around the periphery, **Figure 1(B)**. These tubular aggregates persisted through the remaining stages of the infection, but the vesicles with fibrils decreased markedly in number as the infection progressed, **Figure 2(A)** and **(B)**. Remnant vesicles usually occurred singly, and many were still enclosed in the second membrane.

Abbreviations

AuNPs	Gold nanoparticles
ICP-MS	Inductively coupled plasma mass spectrometry
ER	Endoplasmic reticulum
En	Endosome
AM	Amorphous
F	Filament
VLPs	Virus-like particles
My	Myelin body
CW	Cell Wall

Author details

Noorah Abdulaziz Othman Alkubaisi[1]* and Nagwa Mohammed Amin Aref[2]

1 Department of Botany and Microbiology, College of Science, King Saud University, Riyadh, Kingdom of Saudi Arabia

2 Department of Microbiology, College of Agriculture, Ain Shams University, Cairo, Egypt

*Address all correspondence to: nalkubaisi@ksu.edu.sa

References

[1] I. Amai, M. Kojima, Cytopathic alterations of barley root tissues infected with barley yellow dwarf virus, Bulletin of the Faculty of Agriculture, Niigata University (42) (1990) 25-33.

[2] G. Zhai, KS. Walters, D.W. Peate, P.J. Alvarez, J.L. Schnoor, Transport of gold nanoparticles through plasmodesmata and precipitation of gold ions in woody poplar, Environmental science & technology letters 1(2) (2014) 146-151.

Chapter 6

The Intervention of Gold Nanoparticles (AuNPs) Interactions Lead to the Disappearing of Virus Particles

Noorah Abdulaziz Othman Alkubaisi and Nagwa Mohammed Amin Aref

Abstract

In the context of plant-pathogen interaction, the application of nanoparticle technology and efficient transportation of substances, such as systemic AuNPs to the specific coupling of AuNPs and virus, provide novel solutions for the treatment of plants against the virus. The included data proved that AuNPs provide an efficient means to control virus infection in a fashion way with reducing collateral damage. The AuNPs assure fatal damage to the VLPs with low concentration using different AuNPs sizes. Synergistic therapeutic effects could lead to virus resistance.

Keywords: Barley Yellow Dwarf Virus (BYDV-PAV), Gold nanoparticles (AuNPs)

1. Introduction

In our study by inducing virus particles combining with AuNPs, having an interaction performance (**Figures 1** and **2**), as illustrated in TEM disappearing **Figure 3(B)** or/and swelling **Figure 4(B)**, smashing **Figure 4(C)**, deformation **Figure 4(E)** and **(F)**, corrosive, and puffiness of infected virus particles **Figure 4(I)**, **(G)**, and **(H)** [1]. Multicomponent surface chemistry has also been explored in the context of gene regulation [2]. Modification of the Au NPs with more than one class of functional group can enable multiple functionalities. One example is that Au NPs have been derivatized with both antisense oligonucleotides and thiolated nuclear localization signal (NLS) peptides. The oligonucleotides on the AuNPs surface are close enough. The counter ions associated with one oligonucleotide also act to screen negative charge on adjacent oligonucleotides; this additional charge screening results in more excellent oligonucleotide duplex stabilization relative to free DNA strands. It explains why the DNA AuNPs aggregates melt at higher temperatures than the same unconjugated oligonucleotide duplexes under the same conditions [3].

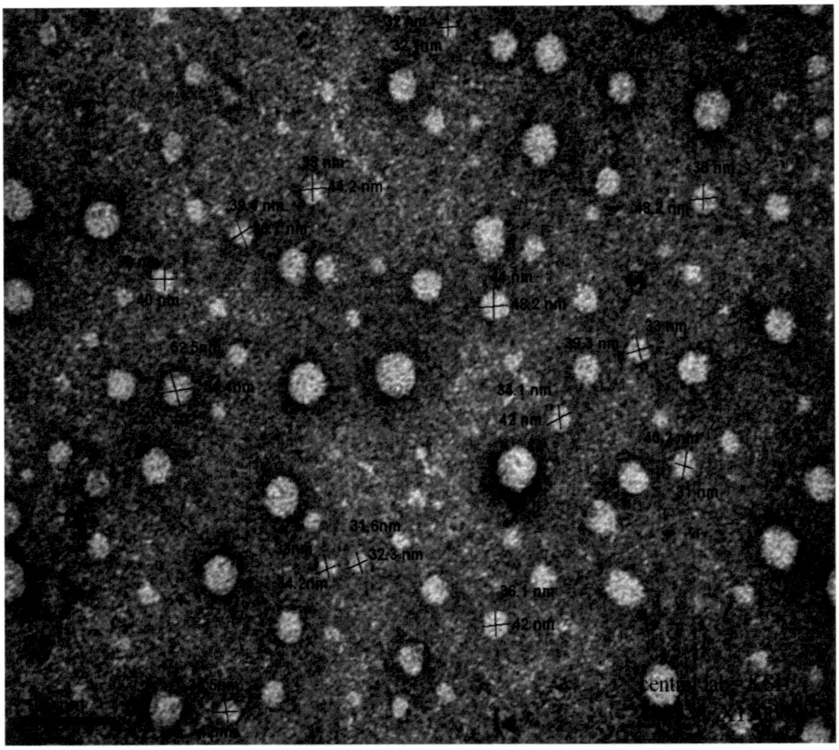

Figure 1.
Electron micrograph showing a purified crude sap of barley yellow dwarf virus (BYDV-PAV), micrograph of BYDV-PAV purification (Control of virus particles). Scale bars 100 nm.

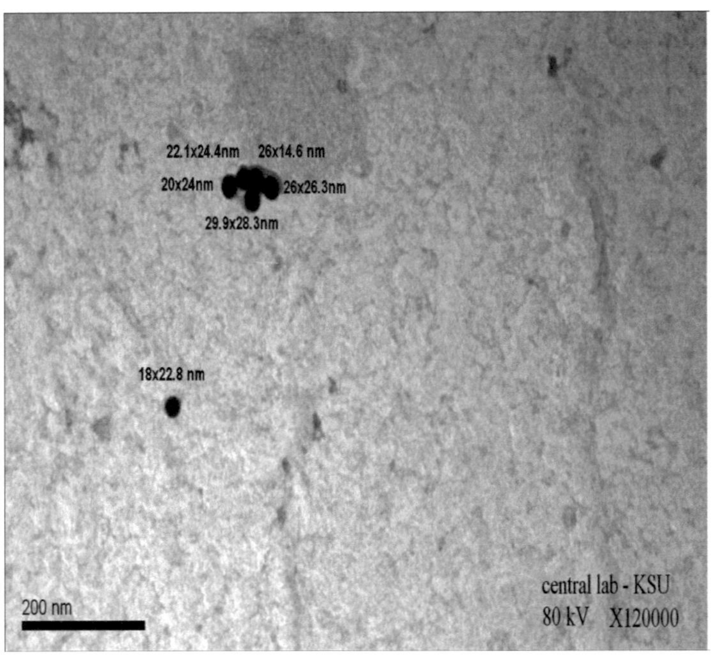

Figure 2.
Electron micrograph showing a purified crude sap of Gold nanoparticles (AuNPs). Micrograph of the size for AuNPs particles (Control of Nanoparticles). Scale bars 200 nm.

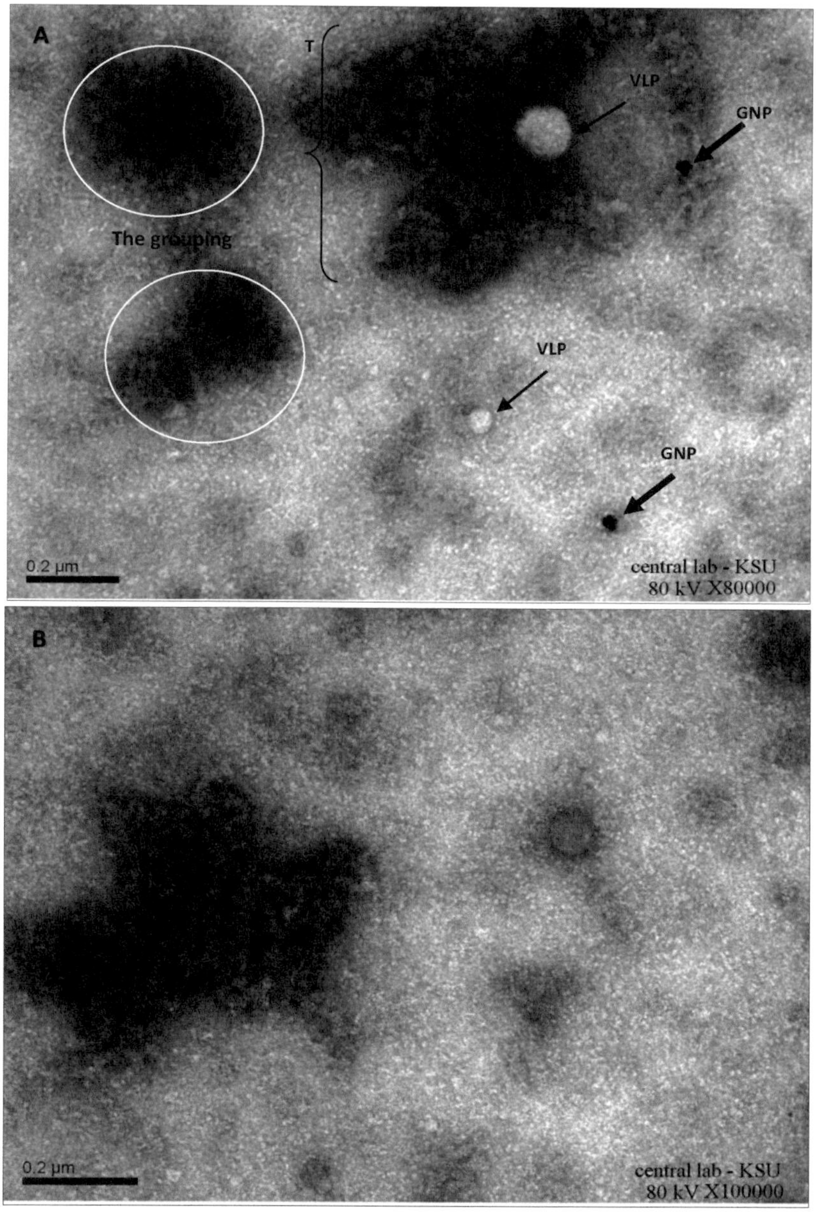

Figure 3.
Electron micrographs showing the treatments of gold nanoparticles (AuNPs) with virus particles In vitro *for 24 hr. (A) Aggregated the virus particles within AuNPs (outlined) with one blowing VLP particle inside the batch's areas of accumulated AuNPs. Scale bar 0.2 µm. (B) Higher magnification of (A). Scale bar 0.2 µm.*

2. Five determinant factors and viral deformation

The application of AuNPs revealed potential damage on VLPs according to five determinant factors that play an essential role in viral particle deformation. First, the incubation period's time duration with virus particles; (24 hr.& 48 hr.) incubation period exhibited deformation highly and vanishing VLPs weather (in *vivo* or in *vitro*). Second & third; Concentration and toxicity of AuNPs. 0.00034 g AuNPs was an initial concentration in one ml distilled water, which

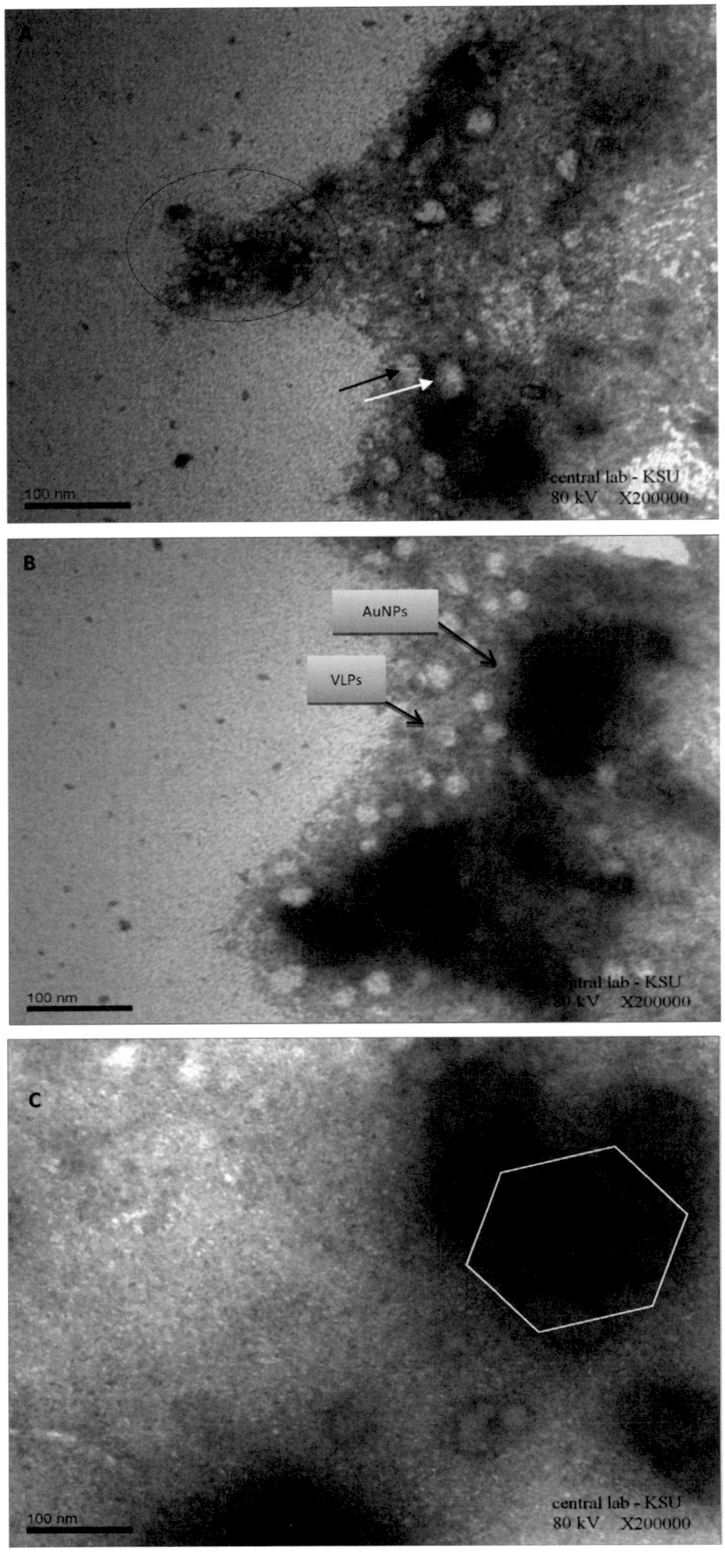
A
central lab - KSU
80 kV X200000
100 nm
B
AuNPs
VLPs
100 nm
C
central lab - KSU
80 kV X200000
100 nm

DOI: http://dx.doi.org/10.5772/intechopen.97443

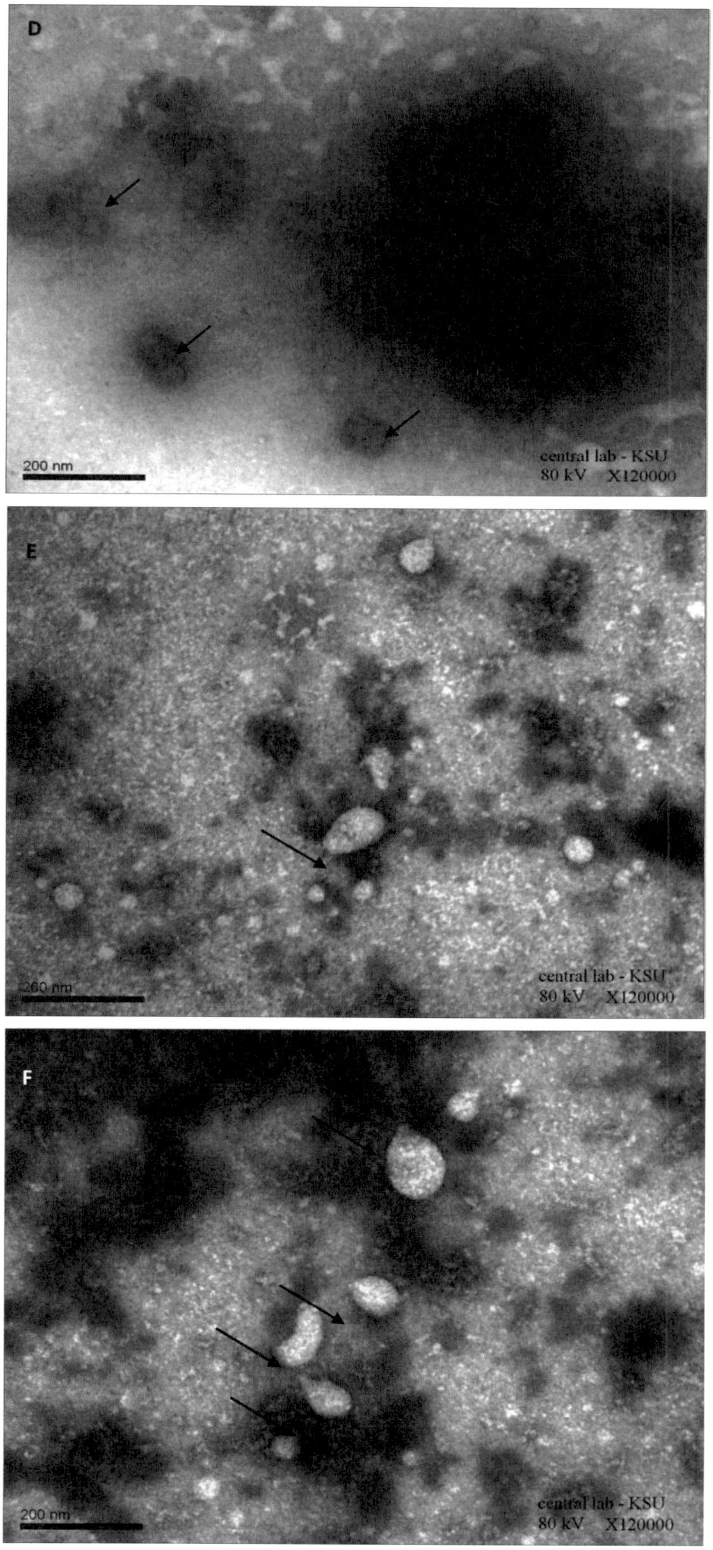

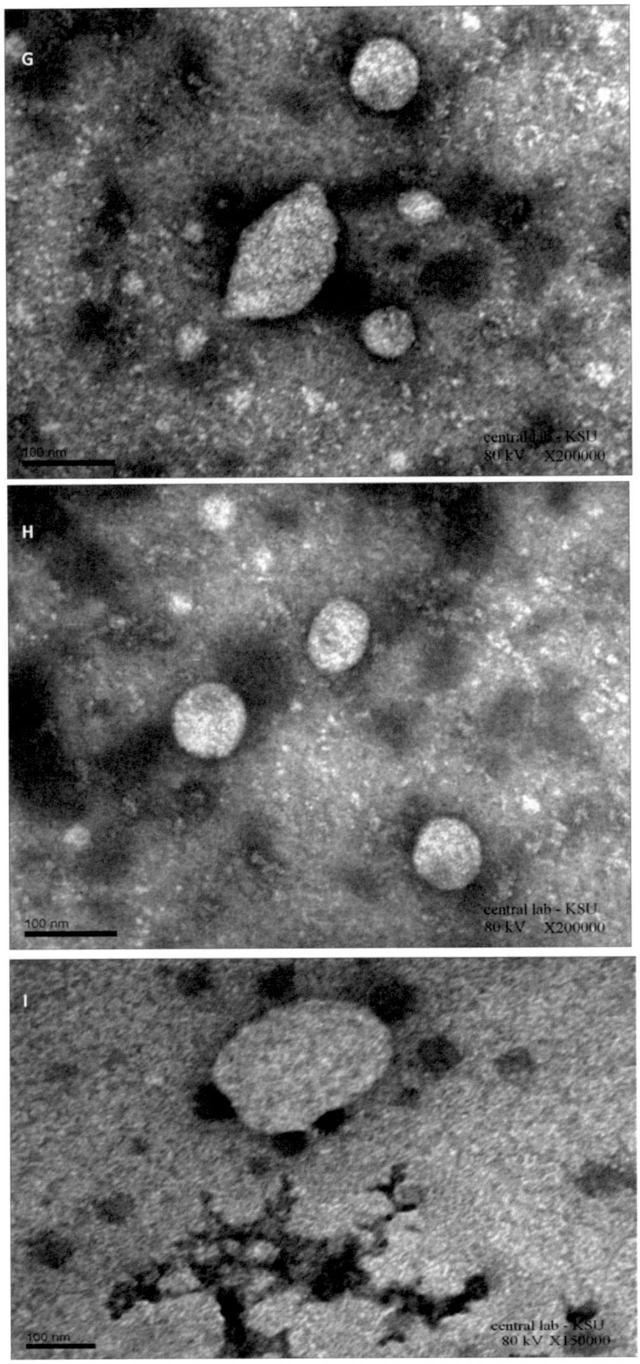

Figure 4.
Electron micrographs showing gold nanoparticles (AuNPs) with virus particles In vivo. *(A) Clotting and gathering VLPs (outlined), and then started to be billowed with dissociated unrounded particles (arrows). Scale bar 100 nm. (B) Some of the VLPs noticed in large size, and AuNPs clotted the particles around it. (arrows). Scale bar 100 nm. (C) Higher magnification for clotting crushed and smashed VLPs (outlined). Scale bar 200 nm. (D) AuNPs surround the VLPs, and the shapes of the particles were damaged and having asymmetric shapes in some other fields (arrows). Scale bar 200 nm. (E) Moreover, (F) high magnification for the deformed, abnormal particles having piriformis shape, different sizes, and shapes of the particles resulting from mixing AuNPs with crude sap as an inoculum (arrows). Scale bars 200 nm. (G) Moreover, (H) higher magnification indicating some individual abnormalities of blowing viral particles. Scale bars 100 nm. (I) Aggregation of VLPs with AuNPs and blown VLP particles surrounded by AuNPs. Scale bar 100 nm.*

had a significant effect on VLPs compared to the diluted one 0.00017 gm of AuNPs. It had excellent detrition results on virus particles in TEM, as shown in **Figures 3** and **4** as an application of AuNPs with virus infection.

Author details

Noorah Abdulaziz Othman Alkubaisi[1*] and Nagwa Mohammed Amin Aref[2]

1 Department of Botany and Microbiology, College of Science, King Saud University, Riyadh, Kingdom of Saudi Arabia

2 Department of Microbiology, College of Agriculture, Ain Shams University, Cairo, Egypt

*Address all correspondence to: nalkubaisi@ksu.edu.sa

References

[1] N.A. Alkubaisi, N.M. Aref, Dispersed gold nanoparticles potentially ruin gold barley yellow dwarf virus and eliminate virus infectivity hazards, Applied Nanoscience 7(1-2) (2017) 31-40.

[2] D.A. Giljohann, D.S. Seferos, W.L. Daniel, M.D. Massich, P.C. Patel, C.A. Mirkin, Gold nanoparticles for biology and medicine, Angewandte Chemie International Edition 49(19) (2010) 3280-3294.

[3] R. Nair, S.H. Varghese, B.G. Nair, T. Maekawa, Y. Yoshida, D.S. Kumar, Nanoparticulate material delivery to plants, Plant science 179(3) (2010) 154-163.